AVIAN INFLUENZA

Wageningen UR Frontis Series

VOLUME 8

Series editor:

R.J. Bogers
*Frontis – Wageningen International Nucleus for Strategic Expertise,
Wageningen University and Research Centre, Wageningen, The Netherlands*

Online version at http://www.wur.nl/frontis

AVIAN INFLUENZA

Prevention and Control

Edited by

REMCO S. SCHRIJVER

*Animal Sciences Group,
Lelystad, Netherlands*

and

G. KOCH

*Central Institute for Disease Control,
Lelystad, Netherlands*

A C.I.P. Catalogue record for this book is available from the Library of Congress.

ISBN-1-4020-3440-7 (PB)
ISBN-978-1-4020-3440-4 (PB)
ISBN-1-4020-3439-3 (HB)
ISBN-978-1-4020-3439-8 (HB)
ISBN-1-4020-3441-1 (e-book)
ISBN-978-1-4020-3441-1 (e-book)

Published by Springer,
P.O. Box 17, 3300 AA Dordrecht, The Netherlands.

www.springeronline.com

Printed on acid-free paper

Contents

Preface

Avian influenza (AI) has become one of the most dangerous infections for animals and humans at present. The outbreaks of the last decade clearly demonstrated that the disease did not limit itself to birds, but that human health was also at stake. The outbreak of avian influenza in Hong Kong in 1997, when several people died following an outbreak in poultry, unfortunately proved not to be an exception to the rule. During the H7N7 AI outbreak in The Netherlands, many cases of conjunctivitis, including one fatal case, occurred, showing that AI may initiate a productive infection cycle with clinical signs also in humans. The classical epidemiological model of avian influenza-virus spread, from waterfowl to poultry to pigs to man, proved to be untrue. Furthermore, mutation of low-pathogenic strains into highly pathogenic strains in poultry appeared to be more common than previously thought. Direct transmission from poultry to man was demonstrated, and even evidence was presented on transmission between humans. These findings suddenly complicated the epidemiological pattern and consequently demanded a complete change in the control policy. However, many questions could not be answered. What was the reason for the increased occurrence? How could the change in transmission patterns be explained? Should low-pathogenic strains in wild birds be monitored? To what extent can vaccination reduce the risks for man and animal? What is the current role of waterfowl and what is the impact of migratory routes? Which strains were dangerous to man? Policymakers, EU officials, researchers, doctors, veterinarians, poultry farmers and industry members are struggling with these questions and at the same time must combat the disease at the best of their knowledge.

These developments have highlighted the need for collaboration between human and veterinary organizations such as the OIE, FAO and WHO. It has also highlighted the need for transparency of countries at risk about the occurrence and control measures of avian influenza. Evidently, avian influenza cannot be controlled blindfolded, but can only be overcome when there is openness about the risk factors and commitment between the control bodies involved.

This led us to organize an international workshop on avian influenza with experts from Europe, the US and Asia. The overall purpose of the AI workshop was to review the AI epidemiology and to identify a research agenda for more effective AI prevention and control in the future. The present volume is based on the papers that were presented during this workshop; a reflection of the discussions can be found on www.wur.nl/frontis.

The ambition was to stimulate progress in AI prevention and control by creating interaction between science, economics and legislation. We hope to have achieved this goal, and thank all the people involved for their contribution. We thank Wageningen University and Research Centre, Intervet International and Merial for their financial support, Petra van Boetzelaer and Rob Bogers (Frontis) for organizational and editorial support, and Paulien van Vredendaal (Wageningen UR Library) for technical editing of these proceedings.

Avian influenza continues to challenge us, but we are ready to face it.

The editors,
Remco Schrijver
Guus Koch

Wageningen, October 2004

INTRODUCTION AND SPREAD OF AVIAN INFLUENZA

1

Avian influenza viruses and influenza in humans

D.J. Alexander[#], I. Capua[##] and I.H. Brown[#]

Abstract

Influenza-A viruses cause natural infections of humans, some other mammals and birds. Few of the 15 haemagglutinin and 9 neuraminidase subtype combinations have been isolated from mammals, but all subtypes have been isolated from birds. There are enormous pools of influenza-A viruses in wild birds, especially migratory waterfowl.

In the 20th century there were 4 pandemics of influenza due to the emergence of antigenically different strains in humans: 1918 (H1N1), 1957 (H2N2), 1968 (H3N2) and 1977 (H1N1). The RNA of influenza-A viruses is segmented into 8 distinct genes and as a result genetic reassortment can occur in mixed infections with different viruses. The 1957 and 1968 pandemic viruses differed from the preceding viruses in humans by the substitution of some genes that came from avian viruses, indicating that pandemic viruses may arise by genetic reassortment of viruses of human and avian origin.

In poultry some influenza-A viruses cause highly pathogenic avian influenza [HPAI], with 100% mortality in infected flocks. The virulence of HPAI viruses is related to the presence of multiple basic amino acids at the precursor HA0 cleavage site, which enables it to be cleaved, and the virus rendered infectious, by a ubiquitous protease (e.g. furin), causing a systemic infection, instead of being restricted to cleavage by trypsin-like proteases. Humans also have furin, but none of the pandemic viruses have had HA0 cleavage sites with multiple basic amino acids.

Up to 1995 there had been only three reports of avian influenza viruses infecting humans, in 1959, 1977 and 1981. All three viruses were H7N7 and two of these infections were the result of laboratory accidents. However, since 1996 there have been regular reports of natural infections of humans with avian influenza viruses. Isolations of avian influenza viruses from humans in England in 1996 [H7N7], Hong Kong in 1997 [H5N1] and 1999 [H9N2] caused concern that a new influenza pandemic could begin. The H5N1 virus was especially alarming, as it possessed multiple basic amino acids at the HA0 cleavage site and 6/18 of the people infected died. In 2003 further human infections with H5N1 virus were reported in Hong Kong with two associated deaths and in The Netherlands a total of 82 people were confirmed as infected with the H7N7 virus responsible for a series of HPAI outbreaks in poultry, one death was reported. Although these infections seem to have been limiting, with very little human to human transmission, they are a cause for alarm since, if people infected with an 'avian' virus were infected simultaneously with a

[#] Virology Department, Veterinary Laboratories Agency Weybridge, Addlestone, Surrey KT15 3NB, United Kingdom. E-mail: d.j.alexander@vla.defra.gsi.gov.uk
[##] Istituto Zooprofilattico Sperimentale delle Venezie, Viale dell'Università 10, 35020 Legnaro, Padova, Italy. E-mail: icapua@izsvenezie.it

1

'human' influenza virus, reassortment could occur with the potential emergence of a virus fully capable of spread in the human population, but with an HA for which the human population was immunologically naive.

Keywords: Avian influenza; human influenza; interspecies transmission; pandemics; reassortment; zoonosis

Introduction

Influenza is a highly contagious, acute illness in humans for which there are recognizable accounts of epidemics dating back to ancient times. Influenza viruses have negative-sense RNA genomes and are placed in the *Orthomyxoviridae* family; they are grouped into types A, B and C on the basis of the antigenic nature of the internal nucleocapsid or the matrix protein, these types are now recognized as genera. Influenza-A viruses infect a large variety of animal species including humans, pigs, horses, sea mammals and birds, occasionally producing devastating pandemics in humans. The two surface glycoproteins of the virus, haemagglutinin (HA) and neuraminidase (NA), are the most important antigens for inducing protective immunity in the host and therefore show the greatest variation. For influenza-A viruses 15 antigenically distinct HA [H1-H15] and 9 NA [N1-N9] subtypes are recognized at present; a virus possesses one HA and one NA subtype, apparently in any combination. Although viruses of relatively few subtype combinations have been isolated from mammalian species, all subtypes, in most combinations, have been isolated from birds.

In the 20th century the sudden emergence of antigenically different strains in humans, termed *antigenic shift*, occurred on 4 occasions, 1918 (H1N1), 1957 (H2N2), 1968 (H3N2) and 1977 (H1N1), resulting in pandemics (Nguyen-Van-Tam and Hampson 2003). In 1957 and 1968 the new viruses completely replaced the previous virus in the human population, but in 1977 this did not occur and currently H3N2 and H1N1 viruses both circulate.

Frequent epidemics have occurred between the pandemics as a result of accumulated point mutations in the prevalent virus leading to gradual antigenic change, termed *antigenic drift*, which in turn results in infections in a proportion of the population that has become immunologically susceptible. The inter-pandemic influenza epidemics may have a considerable impact on a given population as a result of significant mortality, especially amongst the elderly and other vulnerable groups, and the severe economic cost associated with debilitating illness in a large portion of the population. Occasionally the degree of antigenic drift is sufficient that a very large proportion of the population is susceptible and severe epidemics occur with world-wide spread.

By far the worst influenza pandemic for which there are accurate records was the one beginning in 1918. It has been estimated that during the pandemic more than 40 million people died (Nguyen-Van-Tam and Hampson 2003). In well-developed countries such as the USA about 0.5% of the population died, but in some communities in Alaska and the Pacific islands half the population perished. Since 1918 the central theme in the study of human influenza has been to understand how antigenic shift occurs and to predict when and how it will next occur.

Human/avian influenza link

The RNA of influenza viruses is segmented into 8 distinct genes, which code for 10 proteins. Because the viral RNA is segmented and is packed at random, genetic reassortment can occur in mixed infections with different strains of influenza-A viruses. This means that when two viruses infect the same cell, progeny viruses may inherit sets of RNA segments made up of combinations of segments identical to those of either of the parent viruses. This gives a theoretical possible number of 2^8 (=256) different combinations that can form a complete set of RNA segments from a dual infection, although in practice only a few progeny virions possess the correct gene constellation required for viability. Demonstration that the H3N2 1968 pandemic virus differed from the 1957-1968 H2N2 virus in the substitution of two genes, PB1 and the important surface-glycoprotein HA gene, with genes almost certainly from an influenza virus of avian origin, led to the suggestion that antigenic shift occurred as a result of reassortment of genes in dual infections with viruses of human and avian origin (Fang et al. 1981; Gething et al. 1980; Kawaoka, Krauss and Webster 1989; Scholtissek et al. 1978). As a result systematic surveillance studies into the presence of influenza viruses in avian species were undertaken. These revealed enormous pools of influenza-A viruses in wild birds, especially migratory waterfowl. In a series of surveillance studies involving over 20,000 birds during 1973-1986, virus was isolated from about 10%, with an isolation rate of approximately 15% from ducks and geese and 2% from other birds (Alexander 2000). In addition, unlike mammals, where the number of subtypes that have been established appears to be limited, all 15 H and 9 N subtypes recognized currently have been recorded in birds in most possible combinations.

This wealth of influenza viruses in the bird population brought into question the reassortment theory for the origins of pandemic virus, as transfer and reassortment would seem likely to occur much more frequently than subtype changes have appeared in the human population. However, volunteer experiments had shown that only transitory infections resulted when humans were infected with viruses of avian origin (Beare and Webster 1991) and at that time few natural infections of humans with avian viruses had been reported (see below). It was clear that there was some barrier to the establishment of avian influenza viruses in the human population that was related to one or more of the gene segments. Both human and avian viruses are known to infect pigs readily and it was suggested that pigs acted as 'mixing vessels' in which reassortment between human and avian influenza viruses could take place with the emergence of viruses with the necessary gene(s) from the virus of human origin to allow replication and spread in the human population, but with a different haemagglutinin surface glycoprotein, so that the human population could be regarded as immunologically naive. This theory was also thought to account for the apparent emergence of pandemics in the 20th century in the Far East where agricultural practices mean that high concentrations of people, pigs and waterfowl live closely together (Shortridge and Stuart-Harris 1982).

The emergence of pandemic virus may be even more complicated and two hypotheses have been proposed for the rhythm of occurrence of human influenza A viruses, which were termed by Shortridge (1992) as an influenza circle or cycle and an influenza spiral, respectively. The circulation theory suggests there is simply a recycling of H1, H2 and H3 subtypes. If this is so, the HA subtype of the next pandemic virus would be H2. The spiral theory presupposes that humans are capable of being infected with all HA subtypes of influenza-A viruses providing a specific

3

constellation of the other genes is present and that it is a lottery as to which of the 15 recognized will emerge next by reassortment. The theories are not mutually exclusive and since the circulation theory does not state how the next pandemic virus will arise, it could be by reassortment.

Avian influenza pathogenicity

Influenza-A viruses infecting poultry can be divided into two distinct groups on the basis of their ability to cause disease. The very virulent viruses cause a disease formerly known as fowl plague and now termed highly pathogenic avian influenza [HPAI] in which mortality may be as high as 100%. These viruses have been restricted to subtypes H5 and H7, although not all viruses of these subtypes cause HPAI. All other viruses cause a much milder disease consisting primarily of mild respiratory disease, depression and egg-production problems in laying birds. Sometimes other infections or environmental conditions may cause exacerbation of influenza infections leading to much more serious disease.

The main functional glycoprotein, the haemagglutinin, for influenza viruses is produced in a precursor form, HA0, which requires post-translational cleavage by host proteases before the protein is functional and the virus particles are infectious. It has been demonstrated that the HA0 precursor proteins of avian influenza viruses of low virulence for poultry are limited to cleavage by host proteases such as trypsin and trypsin-like enzymes and thus restricted to replication at sites in the host where such enzymes are found, i.e. the respiratory and intestinal tracts. In contrast virulent viruses appear to be cleavable by (a) ubiquitous protease(s), which remains to be fully identified but appears to be one or more proprotein-processing subtilisin-related endoproteases of which furin is the leading candidate (Stieneke-Grober et al. 1992), and this enables these viruses to replicate throughout the animal, damaging vital organs and tissues which brings about disease and death in the infected bird.

Comparisons of the amino-acid sequences at the HA0 cleavage site of avian influenza viruses of high and low pathogenicity revealed that while viruses of low virulence have a single basic amino acid (arginine) at the site, all HPAI viruses possessed multiple basic amino acids (arginine and lysine) adjacent to the cleavage site either as a result of apparent insertion or apparent substitution (Senne et al. 1996; Vey et al. 1992; Wood et al. 1993). The additional basic amino acids result in a motif recognized and cleavable by the putative ubiquitous protease(s). Mammals, including humans, also have furin-like proteases capable of cleaving at multiple basic amino-acid motifs.

Human infections with avian influenza viruses (Table 1)

There were three reports of human infections with avian influenza virus in the literature prior to 1996. In 1959 an HPAI virus of was obtained from a patient with hepatitis (Campbell, Webster and Breese Jr. 1970). The second related to a laboratory worker in Australia who developed conjunctivitis after accidental exposure directly in the eye with a HPAI virus (Taylor and Turner 1977). The third also related to conjunctivitis as the result of infection with an avian LPAI virus, which spread to an animal handler from an infected seal (Webster et al. 1981). Interestingly all three of these viruses were of H7N7 subtype.

In 1996 an H7N7 virus was isolated in England from the eye of a woman with conjunctivitis who kept ducks. This virus was shown to be genetically closest in all 8 genes to viruses of avian origin and to have >98% nucleotide homology in the HA gene with a virus of H7N7 subtype isolated from turkeys in Ireland in 1995 (Banks, Speidel and Alexander 1998).

Table 1. Reports of human infections with avian influenza viruses

Year	Subtype	HPAI/LPAI[1]	Number of people infected	Symptoms
1959	H7N7	HPAI	1	hepatitis?
1977	H7N7	HPAI	1	conjunctivitis
1981	H7N7	LPAI	1	conjunctivitis
1996	H7N7	LPAI	1	conjunctivitis
1997	H5N1	HPAI	18	influenza-like illness, 6 deaths
1999/98	H9N2	LPAI	2 (+5?)	influenza-like illness
2003	H5N1	?	2 (+1?)	influenza-like illness, 1 (+1?) death(s)
2003	H7N7	HPAI	82	conjunctivitis, some cases of influenza-like illness, 1 death

[1]HPAI or LPAI in chickens. See text for source.

In May 1997 a virus of H5N1 subtype was isolated from a young child who died in Hong Kong and by December 1997 the same virus was confirmed by isolation to have infected 18 people, six of whom died (Shortridge et al. 2000). There was evidence of very limited human-to-human spread of this virus (Buxton Bridges et al. 2000), but clearly the efficiency of transmission must have been extremely low. The viruses isolated from the human cases appeared to be identical to viruses first isolated from chickens in Hong Kong in March 1997 following an outbreak of HPAI. Both human and avian isolates possess multiple basic amino acids at the HA0 cleavage site (Suarez et al. 1998).

In recent years outbreaks in poultry due to viruses of H9 subtype, usually H9N2, have been widespread. During the second half of the 1990s outbreaks, due to H9N2 subtype had been reported in Germany, Italy, Ireland, South Africa, USA, Korea, China, the Middle East, Iran and Pakistan (Banks et al. 2000) and this virus continues to spread. In March 1999 two independent isolations of influenza virus subtype H9N2 were made from girls aged one and 4 who recovered from flu-like illnesses in Hong Kong (Peiris et al. 1999a; 1999b). Subsequently, 5 isolations of H9N2 virus from humans on mainland China in August 1998 were reported.

In 2003 an H5N1 virus was isolated from a father and son in Hong Kong who presented with respiratory illness after returning from the Chinese mainland; the father died. A daughter had become ill and died while visiting the Chinese mainland; it is not known if she was infected with H5N1virus. There were reported to be some genetic differences between the 1997 and the 2003 H5N1 viruses (WHO website: http://www.who.int/mediacentre/releases/2003/pr17/en/).

During the 2003 HPAI H7N7 outbreaks in The Netherlands of 260 people involved in some aspect of the outbreak and presenting with conjunctivitis and/or influenza-like illness 82 were confirmed as infected with H7 virus (Koopmans et al. 2003). There was also evidence of three cases of human-to-human transmission within families. Six people tested proved positive for H3N2 influenza, but none were also positive for H7N7. Following these cases all staff involved in the outbreaks were treated prophylactically with antiviral drugs and subjected to vaccination against human influenza (to reduce the chance of reassortment between human and avian viruses).

During this outbreak a human fatality also occurred. The victim was a 57-year-old veterinarian who had not received prophylactic antiviral drugs and had contact with infected birds during outbreak management. He was admitted to hospital with severe headache and fever, subsequently he developed a severe respiratory condition, kidney failure and died. H7 virus was recovered from a broncho-alveolar lavage collected 9 days after the onset of illness (Koopmans et al. 2003).

Five of the eight reports of avian-influenza infections in humans have been with H7N7 subtype viruses. The significance of this is not known.

Conclusions

The high mortality, 6/18, amongst the people infected with the H5N1 virus in Hong Kong was worrying in case the virus was capable of systemic infection due to the presence of multiple basic amino acids at the HA0 cleavage site allowing cleavage to be mediated by (a) furin-like protease(s). However, evidence that this was the case is lacking. Generally, the 18 patients presented with severe respiratory symptoms and for those who died – several of whom were vulnerable due to complicating medical conditions present prior to infection – pneumonia appeared to be the main cause as it often is in deaths occurring as a result of infections with influenza viruses 'normally' in the human population. Similarly the single death amongst those infected with the HPAI H7N7 virus in The Netherlands in 2003 was also the result of pneumonia. Infections of other mammals with avian influenza viruses also give few clues to the significance of multiple basic amino acids at the HA0 cleavage site. An infection of harbour seals during 1978-80 off the Northeast coast of the United States of America with H7N7 avian influenza resulted in death of an estimated >20% of the population. While this mortality rate is comparable to that occurring in humans in Hong Kong, the HA0 cleavage site of the H7N7 virus did not have a motif containing multiple basic amino acids (Webster et al. 1992). Conversely, H7N7 viruses responsible for equine influenza type 1, for which A/equine/Prague/56 is the type strain, do have multiple basic amino acids at the HA0 cleavage site and yet in infections of horses with this strain, virus replication is invariably restricted to the respiratory tract (Gibson et al. 1992).

The demonstration of direct natural infections of humans with avian viruses suggests that pandemic viruses could emerge as a result without an intermediate host. However, for the human population as a whole the main danger is probably not directly the viruses that have spread from avian species, but if the people infected with the avian influenza viruses had been infected simultaneously with a 'human' influenza virus, reassortment could have occurred with the potential emergence of a virus fully capable of spread in the human population, but with H5, H7 or H9 HA, resulting in a true influenza pandemic.

References

Alexander, D.J., 2000. A review of avian influenza in different bird species. *Veterinary Microbiology,* 74 (1/2), 3-13.

Banks, J., Speidel, E. and Alexander, D.J., 1998. Characterisation of an avian influenza A virus isolated from a human: is an intermediate host necessary for the emergence of pandemic influenza viruses? *Archives of Virology,* 143 (4), 781-787.

Banks, J., Speidel, E.C., Harris, P.A., et al., 2000. Phylogenetic analysis of influenza A viruses of H9 haemagglutinin subtype. *Avian Pathology*, 29 (4), 353-360.

Beare, A.S. and Webster, R.G., 1991. Replication of avian influenza viruses in humans. *Archives of Virology*, 119 (1/2), 37-42.

Buxton Bridges, C., Katz, J.M., Seto, W.H., et al., 2000. Risk of influenza A (H5N1) infection among health care workers exposed to patients with influenza A (H5N1), Hong Kong. *Journal of Infectious Diseases*, 181 (1), 344-348.

Campbell, C.H., Webster, R.G. and Breese Jr., S.S., 1970. Fowl plague virus from man. *Journal of Infectious Diseases*, 122 (6), 513-516.

Fang, R., Min Jou, W., Huylebroeck, D., et al., 1981. Complete structure of A/duck/Ukraine/63 influenza hemagglutinin gene: animal virus as progenitor of human H3 Hong Kong 1968 influenza hemagglutinin. *Cell*, 25 (2), 315-323.

Gething, M.J., Bye, J., Skehel, J., et al., 1980. Cloning and DNA sequence of double-stranded copies of haemagglutinin genes from H2 and H3 strains elucidates antigenic shift and drift in human influenza virus. *Nature*, 287 (5780), 301-306.

Gibson, C.A., Daniels, R.S., Oxford, J.S., et al., 1992. Sequence analysis of the equine H7 influenza virus haemagglutinin gene. *Virus Research*, 22 (2), 93-106.

Kawaoka, Y., Krauss, S. and Webster, R.G., 1989. Avian-to-human transmission of the PB1 gene of influenza A viruses in the 1957 and 1968 pandemics. *Journal of Virology*, 63 (11), 4603-4608.

Koopmans, M., Fouchier, R., Wilbrink, B., et al., 2003. Update on human infections with highly pathogenic avian influenza virus A/H7N7 during an outbreak in poultry in the Netherlands. *Eurosurveillance Weekly*, 7 (18). [http://www.eurosurveillance.org/ew/2003/030501.asp#1]

Nguyen-Van-Tam, J.S. and Hampson, A.W., 2003. The epidemiology and clinical impact of pandemic influenza. *Vaccine*, 21 (16), 1762-1768.

Peiris, M., Yam, W.C., Chan, K.H., et al., 1999a. Influenza A H9N2: aspects of laboratory diagnosis. *Journal of Clinical Microbiology*, 37 (10), 3426-3427.

Peiris, M., Yuen, K.Y., Leung, C.W., et al., 1999b. Human infection with influenza H9N2. *Lancet*, 354 (9182), 916-917.

Scholtissek, C., Rohde, W., Von Hoyningen, V., et al., 1978. On the origin of the human influenza virus subtypes H2N2 and H3N2. *Virology*, 87 (1), 13-20.

Senne, D.A., Panigrahy, B., Kawaoka, Y., et al., 1996. Survey of the hemagglutinin (HA) cleavage site sequence of H5 and H7 avian influenza viruses: amino acid sequence at the HA cleavage site as a marker of pathogenicity potential. *Avian Diseases*, 40 (2), 425-437.

Shortridge, K.F., 1992. Pandemic influenza: a zoonosis? *Seminars in Respiratory Infections*, 7 (1), 11-25.

Shortridge, K.F., Gao, P., Guan, Y., et al., 2000. Interspecies transmission of influenza viruses: H5N1 virus and a Hong Kong SAR perspective. *Veterinary Microbiology*, 74 (1/2), 141-147.

Shortridge, K.F. and Stuart-Harris, C.H., 1982. An influenza epicentre? [southern China]. *Lancet*, II (8302), 812-813.

Stieneke-Grober, A., Vey, M., Angliker, H., et al., 1992. Influenza virus hemagglutinin with multibasic cleavage site is activated by furin, a subtilisin-like endoprotease. *Embo Journal*, 11 (7), 2407-2414.

Suarez, D.L., Perdue, M.L., Cox, N., et al., 1998. Comparisons of highly virulent H5N1 influenza A viruses isolated from humans and chickens from Hong Kong. *Journal of Virology*, 72 (8), 6678-6688.

Taylor, H.R. and Turner, A.J., 1977. A case report of fowl plague keratoconjunctivitis. *British Journal of Ophthalmology,* 61 (2), 86-88.

Vey, M., Orlich, M., Adler, S., et al., 1992. Hemagglutinin activation of pathogenic avian influenza viruses of serotype H7 requires the protease recognition motif R-X-K/R-R. *Virology,* 188 (1), 408-413.

Webster, R.G., Bean, W.J., Gorman, O.T., et al., 1992. Evolution and ecology of influenza A viruses. *Microbiology Reviews,* 56 (1), 152-179.

Webster, R.G., Geraci, J., Petursson, G., et al., 1981. Conjunctivitis in human beings caused by influenza A virus of seals. *New England Journal of Medicine,* 304 (15), 911.

Wood, G.W., McCauley, J.W., Bashiruddin, J.B., et al., 1993. Deduced amino acid sequences at the haemagglutinin cleavage site of avian influenza A viruses of H5 and H7 subtypes. *Archives of Virology,* 130 (1/2), 209-217.

2

Avian influenza viruses in Hong Kong: zoonotic considerations

K.F. Shortridge[#]

Abstract

Since the mid-1970s, Hong Kong has functioned as an influenza sentinel post for southern China, a region identified as a hypothetical epicentre for the emergence of pandemic influenza viruses. Nineteen ninety-seven marked the coming-of-age of animal-influenza studies with the recognition in Hong Kong of an incipient pandemic situation brought about by the infection of chicken and humans with an avian influenza H5N1 (H5N1/97) virus. Slaughter of all poultry across the Hong Kong SAR possibly averted a pandemic. Tracking down the source of the H5N1/97 virus to geese and quail and precursor avian H5N1, H9N2 and H6N1 viruses revealed that it was a triple reassortant. This provided a framework for influenza-pandemic preparedness at the baseline avian level, H5N1-like viruses being recognized in chicken in 2001 and twice in 2002; at the human level, H9N2 and H5N1-like isolations were made in 1999 and 2003, respectively.

In contrast to Europe and elsewhere, where outbreaks of disease in chicken (by H5 and H7 subtypes) often follow migratory bird activity in an area and the subsequent detection of low-pathogenic avian influenza (LPAI) and highly pathogenic avian influenza (HPAI) virus forms, in southern China, chicken and other poultry are raised in a permanent, year-round avian-influenza milieu as a consequence of duck-raising practices. Given this long-established milieu, the question is raised whether East-Asian avian influenza viruses comprise a group with a greater propensity for interspecies transmission and potential for pandemicity. The intensification of the poultry industry worldwide in recent years may influence the behaviour of these viruses in the milieu and elsewhere. Clearly, there are scientific and veterinary health needs to redefine the terms LPAI and HPAI. The extent of their applicability in southern China where there is now evidence of H5 and H9 subtype viruses exhibiting swings in avian host range perhaps with 'smouldering virulence' remains to be seen.

With animal influenza now part of the World Health Organization's "Global Agenda on Influenza Surveillance and Control" the time is now opportune for it and its sister organization dealing with animal health, Office International des Epizooties, to strengthen their links to combat collaboratively the serious threat of avian influenza viruses for humans and animals.

Keywords: influenza; H5N1; H9N2; zoonosis; China; Hong Kong; pandemic

---

[#] Dept. of Microbiology, The University of Hong Kong, University Pathology Building, Queen Mary Hospital, Pokfulam Road, Hong Kong SAR, China. E-mail: kennedyfs@xtra.co.nz

R. S. Schrijver and G. Koch (eds.), Avian Influenza, 9–18.
© 2005 *Springer. Printed in the Netherlands.*

Chapter 2

Hong Kong situation

The import of the term 'influenza' in the public's perception was ratcheted up a number of notches in 1997 when a novel, highly pathogenic influenza-A virus of avian origin, H5N1 (H5N1/97), killed chicken on farms and in live-poultry markets in the Hong Kong SAR as well as six of 18 people known to have been infected (Claas et al. 1998; Subbarao et al. 1997; Yuen et al. 1998). Ominously, a purely avian influenza virus was causing respiratory disease and death in humans. An important source of food was also the source of a zoonotic infectious disease that stood to become a global health threat. Moreover, the viability of this food source globally stood to be affected (Shortridge, Peiris and Guan 2003).

Termed locally as 'bird flu', this avian influenza (AI) outbreak put Hong Kong into an incipient influenza pandemic situation, and a pandemic was probably averted by the slaughter of all chicken and other poultry across the SAR (Shortridge et al. 2000). By building a profile of influenza ecology over the years particularly as it applied to domestic animals in southern China, Hong Kong essentially functioned as an influenza sentinel post for it and the wider region. This led to an element of influenza-pandemic preparedness in animals and humans (Shortridge 1988; 1992). The H5N1 or 'bird flu' incident upheld the hypotheses that southern China is an epicentre for the emergence of pandemic influenza viruses (Shortridge and Stuart-Harris 1982) and that pandemic influenza is a zoonosis (Shortridge 1992; Webster et al. 1992). The latter hypothesis was engendered from the observation that the surface haemagglutinins (HAs) of the 1968 H3N2 (Hong Kong) pandemic influenza virus and an avian influenza virus were antigenically related (Webster and Laver 1975).

The worth of this understanding in the post-1997 period might be measured by the detection of genotypes of H5N1-like viruses in asymptomatically infected chicken in markets in 2001 before they showed signs of disease. Poultry were killed market-by-market as signs became evident, leading to the pre-emptive slaughter of all poultry to prevent human infection (Guan et al. 2002). Early detection and reaction was the order again in 2002 and 2003 (Secretary for the Environment and Food 2002; Guan et al. 2003). Thus, there now lay the prospect for influenza-pandemic preparedness not only at the human level but, better still, at the baseline avian level with the ideal that if a virus could be stamped out before it infected humans, an influenza incident or pandemic will not result (Shortridge, Peiris and Guan 2003). In 1997, the world was probably one or two mutational events away from a pandemic while in 2002, with earlier detection, it was probably three or four events away.

Preparedness at the human level was put to the test in an unexpected manner in early 2003 around the start of the SARS (severe acute respiratory syndrome) outbreak in Hong Kong (Peiris et al. 2003). Two Hong Kong residents, recently returned from Fujian Province in southern China, exhibited illness with a clinical picture not dissimilar from that of SARS. They did not have SARS; instead, H5N1 viruses were isolated from them, one of whom died. A third family member died earlier in Fujian Province from acute respiratory disease that was not subject to laboratory examination (World Health Organization 2003; Guan et al. 2003). No doubt, the paths for preparedness and reaction will continue to be a learning experience. Having low-cost, widely available, reliable diagnostic tests for distinguishing infection by H5N1 virus and the SARS coronavirus is of paramount importance (Shortridge 2003a).

Nineteen ninety-seven might be considered something of a watershed in dealing with a pandemic. Given that the interpandemic periods in the 20th century ranged from 20 years for the probable H3-like virus that straddled the two centuries from

1898 to 1918 (Masurel and Marine 1973), 39 years for the H1N1 virus and 11 years for the H2N2 (Asian) virus and, as the H3N2 (Hong Kong) virus era was 29 years on in 1997, the time was apposite for a new pandemic virus to emerge. The H5N1 virus did emerge in 1997, in doing so dispelling the notion of a dependency on recycling of haemagglutinin (H) subtypes for pandemicity (Shortridge 1992). However, the virus lacked significant human-to-human transmissibility. It might have acquired this trait through reassorting with a prevailing human influenza-A virus in a human (more likely) or porcine (less likely) 'mixing vessel' but the slaughter of poultry across the SAR denied it this possibility (Shortridge et al. 2000).

Changing virus and behaviour

The genotype of the 2003 human isolates was similar to that of a novel H5N1 virus isolated from resident waterfowl and migratory birds in two Hong Kong parks in late 2002, highlighting the potential of such birds for spreading H5N1 viruses within and beyond the influenza epicentre (Sturm-Ramirez et al. 2003). As well, antigenic analysis of the HAs of the new isolates showed that they differed considerably from those of H5N1 viruses of 1997 and 2001, potentially exacerbating an already tricky situation with H5N1 vaccine preparedness efforts. In the avian influenza-virus milieu of southern China, infection of land-based poultry such as quail by influenza viruses from domestic ducks could facilitate the generation of novel variants capable of further interspecies transmission (Perez et al. 2003; Li et al. 2003). Here, the quail as a minor poultry, possibly acts as an intermediate host or as an avian 'mixing vessel' for the generation of variants capable of infecting land-based chicken and other gallinaceous birds, in the process leading to antigenic variation of the HA. The antigenic differences seen in the 2002 and 2003 H5N1 isolates could represent antigenic drift of the H5 HA. The demonstration that HA escape mutants of an avian H5N2 virus can be selected experimentally with monoclonal antibodies is a pointer of the potential of the H5 HA to undergo antigenic drift (Kaverin et al. 2002). This emerging picture about the H5 subtype is relevant to, say, that of the commonly encountered avian H3 subtype that gave rise to the H3N2 pandemic virus of 1968. This subtype exists in domestic ducks in southern China as a range of established antigenic variants each seemingly with its own potential hierarchic capability of infecting the human host and undergoing antigenic drift in it (Shortridge, Underwood and King 1990). As well, the potential for influenza-A viruses to undergo recombination should not be overlooked (Worobey et al. 2002; Chare, Gould and Holmes 2003). The isolation of avian H7N3 viruses from chicken in Chile that had apparently converted from low-pathogenic avian influenza (LPAI) to highly pathogenic avian influenza (HPAI) forms by recombination is of interest here (Suarez et al. 2003).

As part of a momentum toward improved infectious disease intelligence, of which influenza pandemic preparedness is an intrinsic constituent, there is much to be gained by (1) carrying out extensive influenza-virus surveillance to have a better picture of the influenza viruses in nature with particular reference to domestic animals as the most immediate source of avian influenza viruses for humans and (2) conducting detailed antigenic analyses on the HAs of isolates (in conjunction with receptor-binding studies) for insight into the H subtypes and their variants likely to cross the species barrier to humans and undergo 'prolonged' antigenic drift in humans. It is a moot point whether such studies should lead genotyping and molecular analysis, much the norm nowadays, rather than the other way around. This view does not override the pressing need for better understanding of the signatures in the genes that facilitate interspecies transmission. Here, there is much scope for fruitful international collaborative and co-operative studies in the spirit of those undertaken in the SARS outbreak (Stöhr 2003b). There is urgency in this, for each year brings us closer to the next influenza pandemic (Shortridge 1995) bearing in mind that, at this point in influenza history, pandemic influenza is a non-eradicable zoonosis (Shortridge 1992; Webster et al. 1992).

The H5N1 incident in Hong Kong involving, as it did, chicken, might be viewed as part of a changing global pattern that had been building up in recent years, particularly with H5 and H7 subtype viruses, following transitions from non-pathogenicity to LPAI and to HPAI viruses. In the last 100 years or so, outbreaks of disease originally described in the 19th century as 'fowl plague' because of their widespread occurrence and severity, but now known as HPAI, were becoming increasingly rare. In the period 1959-1999, there were 18 outbreaks. However, from 1999, there have been seven major outbreaks involving H5 or H7 viruses (excluding those from Mexico and Pakistan) with around 38 million birds affected (Scientific Committee on Animal Health and Animal Welfare 2000; Capua and Alexander 2003). Apparently non-pathogenic H5 viruses had been isolated in Hong Kong from ducks and a goose in a study some 20 years prior to 1997; isolation of influenza viruses from chicken was rare (Shortridge 1992). Could the H5N1/97 virus incident and follow-on H5N1 events in Hong Kong be the East-Asian dimension of an H5/H7 AI continuum? The isolation of H5N1 HPAI viruses from geese in northern Vietnam in 2001 as part of the wider region of East Asia and their genetic relatedness to recent human and avian H5N1 isolates in Hong Kong, are therefore of as much interest as they are of concern (Nguyen et al. 2003). Could transmission of H7N7 HPAI viruses from poultry to humans in The Netherlands in 2003 giving rise to conjunctivitis and a case fatality (Osterhaus et al. 2003) be a signal of a further interspecies transmissions of H7 virus to humans?

The Hong Kong and Dutch incidents caused by H5N1 and H7N7 HPAI viruses, respectively, show a link between high pathogenicity in chicken with ability to cause disease in humans possibility leading to pandemic scenarios. Attractive as this lead is to a virus's potential for pandemicity, it may be misleading in the sense that the H2N2 and H3N2 pandemic viruses of 1957 and 1968, respectively, lacking significant basic amino-acid motifs at the cleavage points of their HAs, are unlikely to have been pathogenic for chicken. The factors giving rise to each pandemic virus may be different, pandemicity probably being a polygenic trait (Webster et al. 1992). These factors are really not known although there is evidence that the PB1 gene may be a facilitator gene in this (Kawaoka, Krauss and Webster 1989; Lin et al. 1994). To the present, the geographic source of viruses appears to be important. Avian influenza viruses may be divided into two lineages: (1) Eurasian including African and Australasian viruses and (2) North-American, although it remains to be see whether there is a distinct South-American clade (Suarez et al. 2004). These two broad groupings presumably reflect migratory bird routings and virus seedings. In keeping with the historical record and modern genetic information, it might be reasonable to expect avian viruses of the wider East-Asian region encompassing as it does southern China to have, or be capable of having, the necessary factors or signatures for pandemicity. Genetic analysis of avian isolates from southern China is a pointer toward this (Lin et al. 1994). However, insufficient viruses have been genotyped to know whether separate European and Asian sublineages exist with the Eurasian lineage and whether one or other or both hold the secrets of pandemicity. Exploration of this would be a good move within the wider ambit of infectious-disease intelligence and, optimistically, an additional step toward baseline influenza-pandemic preparedness.

The intensification of the poultry industry worldwide seems to be a key element in causing influenza viruses of aquatic origin to undergo 'more rapid' adaptation to land-based poultry as proffered earlier. This has been aptly demonstrated with H9 viruses from southern China whose HAs retain amino-acid signatures compatible with a receptor-binding preference for human tissue (Perez et al. 2003; Li et al. 2003). The spread of H9N2 viruses to almost panzootic proportions (Alexander 2001) is not only a threat to the poultry industry worldwide but a threat to human health. The ability of H9N2 viruses to infect and cause respiratory illness in humans in Hong Kong and China (Peiris et al. 1999; Guo et al. 1999) against the background that H9N2 viruses isolated from poultry in Hong Kong have a cell receptor specificity similar to that of human H3N2 viruses (Matrosovich, Krauss and Webster 2001) underlines the importance of this virus subtype as a potential pandemic virus (Shortridge et al. 2001). The co-circulation of H9N2 viruses and human H3N2 variants in pigs, i.e. the porcine mixing vessel, in southern China takes this potential a step forward (Peiris et al. 2002).

China

What is special about China as a source of epidemics and pandemics in times past (Potter 1998), the events of the last century including H5N1 and H9N2 focusing attention on southern China? Quite simply, the region has a permanent gene pool of avian influenza viruses year-round (Shortridge 1992). This is as a consequence of the domestication of the duck as a source of these viruses around 4500 years ago in the fertile southeastern region of the country and subsequent intensification and spread of duck raising as an adjunct to rice farming around the start of the Ching Dynasty in 1644 A.D. (Needham 1986). The high human population density in the countryside and preference of the population generally for its food to be as fresh as possible have provided on-going opportunities for human (and porcine) exposure to avian influenza viruses (Shortridge and Stuart-Harris 1982; Shortridge 1992; 2003b). The influenza gene pool in southern China is already established and, more than likely, is not dependent upon introductions from migratory birds. Obviously, introductions could re-vitalize it from time to time and place to place, but movements within southern China or the wider region would help to maintain the pool indigenously. Interestingly, virus isolations from migratory birds at a wetland migratory-bird resting point at the edge of Hong Kong over a number of years yielded few isolates (unpublished data) whereas birds in a roughly parallel route through Taiwan to the East yielded a diverse range of virus subtypes (Cheng et al. 2003). By contrast, an avian influenza gene pool similar to the one in southern China does not exist anywhere else in the world. Virus introductions into poultry are usually the result of the presence of migratory waterfowl in the vicinity and subsequent infection of chicken and turkey, especially if the turkey are raised in the open (Alexander 2001). It is worth noting here that, while the avian-influenza milieu in southern China is complex, it could become even more complex should turkey be raised there commercially.

What is not known about the avian-influenza milieu of southern China is whether there have been outbreaks of AI in chicken and other gallinaceous birds given that up to the last 20 or so years, poultry were scattered widely across the region in villages and small holdings. More recently, larger operations have come on-stream for an increasing population particularly in the cities and for export. Due to the limitations of virus surveillance, it is conjectural whether the H5N1/97 virus in Hong Kong would have existed as an LPAI virus before converting to an HPAI virus in the manner of AI outbreaks in chicken in Europe and elsewhere (Osterhaus et al. 2003; Campitelli et al. 2003). As a triple reassortant involving precursor H5N1, H9N2 and H6N1 viruses and goose and quail hosts (Guan et al. 2002) the possibility that the H5N1/97 virus was an HPAI virus at the outset cannot be excluded. The recognition that the precursor H5N1 virus was associated with the death of geese on a farm in Guangdong Province prior to the 1997 incident (Tang et al. 1998; Zu et al. 1999) and in late 2002 H5N1-like viruses were isolated from dead, resident waterfowl and migratory birds in two Hong Kong parks (Sturm-Ramirez et al. 2003) suggests that a range of genotypes and antigenic variants of H5N1 viruses may be lurking in the region with 'smouldering virulence'. A similar, but more insidious situation may be brewing with certain H9N2 viruses following adaptation from aquatic to land-based poultry and back again as double or triple reassortants (Perez et al. 2003; Li et al. 2003). The fact that these viruses are able to retain a receptor-binding specificity for human tissue while exhibiting limited or no pathogenicity for poultry means that their hypothetical 'smouldering virulence' could go unnoticed as could an incipient pandemic situation. Thus, while there is an obvious need to clarify the definitions of LPAI and HPAI for H5 and H7 virus subtypes as well as taking on board pathogenicity changes in other subtypes (Scientific Committee on Animal Health and Animal Welfare 2000), the situation with the influenza viruses in southern China seems to be far more complex in some cases for clear-cut definitions.

Comment

This meeting was a timely recognition of the need to thread a path of understanding through the complex maze of influenza epidemiology and ecology and its impact on the poultry industry. This has a value-added effect on human health in preparedness for influenza pandemics. Appreciation of the animal dimension of human influenza came of age with the H5N1 incident in Hong Kong in 1997. In concordance with these developments, the World Health Organization (WHO) set about establishing an Animal Influenza Network (AIN) through an inaugural meeting in Hong Kong in 2000 taking as its model the highly successful WHO Influenza Surveillance Network with a view to expanding animal-influenza surveillance in a more co-ordinated fashion and integrating it with human-influenza surveillance. Animal influenza formally took its place in the wide ranging "Global Agenda on Influenza Surveillance and Control" at a "WHO Consultation on Global Priorities in Influenza Surveillance and Control" in Geneva, May 2002 (Stöhr 2003a). An important component of the animal agenda is the need to establish close links between the Office International des Epizooties (OIE), the world organization that oversees animal health, and WHO networks. In doing so, it is highlighting the desirability, if not the moral responsibility, of nations to make available to the best of their circumstances information on influenza and its viruses for the common good of man and animal.

References

Alexander, D.J., 2001. Ecology of avian influenza in domestic birds. *In:* Vicari, M. ed. *Emergence and control of zoonotic ortho- and paramyxovirus diseases.* Libbey Eurotext, Paris, 25-33.

Campitelli, L., Mogavero, E., De Marco, M.A., et al., 2003. Identification in the wild bird reservoir of virus precursors of H7N3 influenza strains responsible for low pathogenic avian influenza in domestic poultry in Italy during 2002-03 [abstract]. *In:* Kawaoka, Y. ed. *Options for the control of influenza V: proceedings of the international conference for the control of influenza V, Okinawa, Japan, October 7-11, 2003.* Elsevier, Amsterdam, 125. International Congress Series no. 1263.

Capua, I. and Alexander, D.J., 2003. Recent developments on avian influenza [abstract]. *In:* Kawaoka, Y. ed. *Options for the control of influenza V: proceedings of the international conference for the control of influenza V, Okinawa, Japan, October 7-11, 2003.* Elsevier, Amsterdam, 123. International Congress Series no. 1263.

Chare, E.R., Gould, E.A. and Holmes, E.C., 2003. Phylogenetic analysis reveals a low rate of homologous recombination in negative-sense RNA viruses. *Journal of General Virology,* 84 (10), 2691-2703.

Cheng, M.C., Lee, M.S., Wang, C.H., et al., 2003. Influenza A virological surveillance in migratory waterfowl in Taiwan from 1998 to 2002 [abstract]. *In:* Kawaoka, Y. ed. *Options for the control of influenza V: proceedings of the international conference for the control of influenza V, Okinawa, Japan, October 7-11, 2003.* Elsevier, Amsterdam, 123. International Congress Series no. 1263.

Claas, E.C.J., Osterhaus, A.D.M.E., Van Beek, R., et al., 1998. Human influenza A H5N1 virus related to a highly pathogenic avian influenza virus. *Lancet,* 351 (9101), 472-477.

Guan, Y., Peiris, J.S.M., Lipatov, A.S., et al., 2002. Emergence of multiple genotypes of H5N1 avian influenza viruses in Hong Kong SAR. *Proceedings of the National Academy of Sciences of the United States of America,* 99 (13), 8950-8955.

Guan, Y., Poon, L.L.M., Yuen, K.Y., et al., 2003. The re-emergence of H5N1 influenza virus in humans: a renewal of pandemic concern? [abstract]. *In:* Kawaoka, Y. ed. *Options for the control of influenza V: proceedings of the international conference for the control of influenza V, Okinawa, Japan, October 7-11, 2003.* Elsevier, Amsterdam, 52-53. International Congress Series no. 1263.

Guo, Y.J., Li, J.W., Cheng, X., et al., 1999. Discovery of humans infected by avian influenza A (H9N2) virus. *Chinese Journal of Experimental and Clinical Virology,* 13, 105-108.

Kaverin, N.V., Rudneva, I.A., Ilyushina, N.A., et al., 2002. Structure of antigenic sites on the haemagglutinin molecule of H5 avian influenza virus and phenotypic variation of escape mutants. *Journal of General Virology,* 83 (10), 2497-2505.

Kawaoka, Y., Krauss, S. and Webster, R.G., 1989. Avian-to-human transmission of the PB1 gene of influenza A viruses in the 1957 and 1968 pandemics. *Journal of Virology,* 63 (11), 4603-4608.

Li, K.S., Xu, K.M., Peiris, J.S.M., et al., 2003. Characterization of H9 subtype influenza viruses from the ducks of Southern China: a candidate for the next influenza pandemic in humans? *Journal of Virology,* 77 (12), 6988-6994.

Lin, Y.P., Shu, L.L., Wright, S., et al., 1994. Analysis of the influenza virus gene pool of avian species from Southern China. *Virology,* 198 (2), 557-566.

Masurel, N. and Marine, W.M., 1973. Recycling of Asian and Hong Kong influenza A virus hemagglutinins in man. *American Journal of Epidemiology,* 97 (1), 44-49.

Matrosovich, M.N., Krauss, S. and Webster, R.G., 2001. H9N2 influenza A viruses from poultry in Asia have human virus-like receptor specificity. *Virology,* 281 (2), 156-162.

Needham, J., 1986. Biological pest control. *In:* Hsing-Tsung, H. ed. *Science and civilization in China. Vol. 6. Biology and biological technology. Part 1. Botany.* Cambridge University Press, Cambridge, 519-553.

Nguyen, D.C., Uyeki, T., Jadhao, S.J., et al., 2003. Avian influenza viruses, including highly pathogenic H5N1, circulate in live poultry in northern Vietnam [abstract]. *In:* Kawaoka, Y. ed. *Options for the control of influenza V: proceedings of the international conference for the control of influenza V, Okinawa, Japan, October 7-11, 2003.* Elsevier, Amsterdam, 125. International Congress Series no. 1263.

Osterhaus, A., Kuiken, T., Munster, V., et al., 2003. HPAI H7N7 in The Netherlands: wild birds, poultry and humans [abstract]. *In:* Kawaoka, Y. ed. *Options for the control of influenza V: proceedings of the international conference for the control of influenza V, Okinawa, Japan, October 7-11, 2003.* Elsevier, Amsterdam, 124. International Congress Series no. 1263.

Peiris, J.S., Lai, S.T., Poon, L.L., et al., 2003. Coronavirus as a possible cause of severe acute respiratory syndrome. *Lancet,* 361 (9366), 1319-1325.

Peiris, J.S.M., Guan, Y., Markwell, D., et al., 2002. Co-circulation of avian H9N2 and contemporary "human" H3N2 influenza viruses in pigs in southern China: potential for genetic reassortment? *Journal of Virology,* 75 (20), 9679-9686.

Peiris, M., Yuen, K.Y., Leung, C.W., et al., 1999. Human infection with influenza H9N2. *Lancet,* 354 (9182), 916-917.

Perez, D. R., Lim, W., Seiler, J.P., et al., 2003. Role of quail in the interspecies transmission of H9 influenza A viruses: molecular changes on HA that correspond to adaptation from ducks to chickens. *Journal of Virology,* 77 (5), 3148-3156.

Potter, C.W., 1998. Chronicle of influenza pandemics. *In:* Hay, A.J. ed. *Textbook of influenza.* Blackwell Science, Oxford, 3-18.

Scientific Committee on Animal Health and Animal Welfare, 2000. *The definition of avian influenza and The use of vaccination against avian influenza.* European Commission, Scientific Committee on Animal Health and Animal Welfare. [http://europa.eu.int/comm/food/fs/sc/scah/out45_en.pdf]

Secretary for the Environment and Food, 2002. *Report of the investigation team for the 2002 avian influenza incident.* Government of the Hong Kong SAR.

Shortridge, K.F., 1988. Pandemic influenza: a blueprint for control at source. *Chinese Journal of Experimental and Clinical Virology,* 2, 75-89.

Shortridge, K.F., 1992. Pandemic influenza: a zoonosis? *Seminars in Respiratory Infections,* 7 (1), 11-25.

Shortridge, K.F., 1995. The next pandemic influenza virus? *Lancet,* 346 (8984), 1210-1212.

Shortridge, K.F., 2003a. SARS exposed, pandemic influenza lurks. *Lancet,* 361 (9369), 1649.

Shortridge, K.F., 2003b. Severe acute respiratory syndrome and influenza: virus incursions from southern China. *American Journal of Respiratory and Critical Care Medicine,* 168 (12), 1416-1420.

Shortridge, K.F., Gao, P., Guan, Y., et al., 2000. Interspecies transmission of influenza viruses: H5N1 virus and a Hong Kong SAR perspective. *Veterinary Microbiology*, 74 (1/2), 141-147.

Shortridge, K.F., Peiris, J.S. and Guan, Y., 2003. The next influenza pandemic: lessons from Hong Kong. *Journal of Applied Microbiology*, 94 (suppl.), 70S-79S.

Shortridge, K.F., Peiris, M., Guan, Y., et al., 2001. H5N1: beaten but is it vanquished? *In:* Vicari, M. ed. *Emergence and control of zoonotic ortho- and paramyxovirus diseases*. Libbey Eurotext, Paris, 91-97.

Shortridge, K.F. and Stuart-Harris, C.H., 1982. An influenza epicentre? [southern China]. *Lancet*, II (8302), 812-813.

Shortridge, K.F., Underwood, P.A. and King, A.P., 1990. Antigenic stability of H3 influenza viruses in the domestic duck population of southern China. *Archives of Virology*, 114 (1/2), 121-136.

Stöhr, K., 2003a. The global agenda on influenza surveillance and control. *Vaccine*, 21 (16), 1744-1748.

Stöhr, K., 2003b. A multicentre collaboration to investigate the cause of severe acute respiratory syndrome. *Lancet*, 361 (9370), 1730-1733.

Sturm-Ramirez, K.M., Guan, Y., Peiris, M., et al., 2003. H5N1 influenza A viruses from 2002 are highly pathogenic in waterfowl [abstract]. *In:* Kawaoka, Y. ed. *Options for the control of influenza V: proceedings of the international conference for the control of influenza V, Okinawa, Japan, October 7-11, 2003*. Elsevier, Amsterdam, 55. International Congress Series no. 1263.

Suarez, D. L., Senne, D.A., Banks, J., et al., 2003. A virulence shift in the influenza A subtype H7N3 virus responsible for a natural outbreak of avian influenza in Chile appears to be the result of recombination [abstract]. *In:* Kawaoka, Y. ed. *Options for the control of influenza V: proceedings of the international conference for the control of influenza V, Okinawa, Japan, October 7-11, 2003*. Elsevier, Amsterdam, 55. International Congress Series no. 1263.

Suarez, D.L., Senne, D.A., Banks, J., et al., 2004. Recombination resulting in virulence shift in avian influenza outbreak, Chile. *Emerging Infectious Diseases*, 10 (4), 693-699.

Subbarao, K., Klimov, A., Katz, J., et al., 1997. Characterization of an avian influenza A (H5N1) virus isolated from a child with a fatal respiratory illness. *Science*, 279 (5349), 393-396.

Tang, X., Tian, G., Zhao, J., et al., 1998. Isolation and characterization of prevalent strains of avian influenza viruses in China. *Chinese Journal of Animal and Poultry Infectious Diseases*, 20, 1-5.

Webster, R.G., Bean, W.J., Gorman, O.T., et al., 1992. Evolution and ecology of influenza A viruses. *Microbiology Reviews*, 56 (1), 152-179.

Webster, R.G. and Laver, W.G., 1975. Antigenic variation of influenza viruses. *In:* Kilbourne, E.D. ed. *Influenza viruses and influenza*. Academic Press, New York, 269-314.

World Health Organization, 2003. *Influenza A(H5N1) in Hong Kong Special Administrative Region of China – update 2*. World Health Organization. Disease Outbreak News. [http://www.who.int/csr/don/2003_02_27a/en/]

Worobey, M., Rambaut, A., Pybus, O.G., et al., 2002. Questioning the evidence for genetic recombination in the 1918 "Spanish flu" virus. *Science*, 296 (5566), 211 discussion 211.

17

Yuen, K.Y., Chan, P.K., Peiris, M., et al., 1998. Clinical features and rapid viral diagnosis of human disease associated with avian influenza A H5N1 virus. *Lancet,* 351 (9101), 467-471.

Zu, X., Subbarao, K., Cox, N.J., et al., 1999. Genetic characterization of the pathogenic influenza A/Goose/Guangdong/1/96 (H5N1) virus: similarity of its hemagglutinin gene to those of H5N1 viruses from the 1997 outbreaks in Hong Kong. *Virology,* 261 (1), 15-19.

3

Live-bird markets in the Northeastern United States: a source of avian influenza in commercial poultry

D.A. Senne[#], J.C. Pedersen[#] and B. Panigrahy[#]

Abstract

In 1994, an H7N2 subtype avian influenza virus of low pathogenicity was detected in live-bird markets (LBMs) of the Northeast United States. Since that time the H7N2 virus continues to circulate in the LBMs despite efforts to eradicate the virus by market closures followed by extensive cleaning and disinfection. Since 1996, the LBMs have been implicated as the source of virus in five outbreaks of H7N2 avian influenza in commercial poultry. Although the H7N2 virus is of low pathogenicity, several mutations have occurred at, or near, the cleavage site of the haemagglutinin (H) protein, a region of the protein known to influence pathogenicity of H5 and H7 avian influenza viruses. From 1994 to 2002, the amino-acid motif at the H cleavage site has gradually changed from PENPKTR/GLF to PEKPKKR/GLF, with the addition of two lysine (K) residues. Also, a 24-nucleotide deletion, believed to be part of the receptor-binding region, was first observed in LBM H7N2 isolates in 1996 and is seen in all isolates tested since 2000. These findings support the need to continue avian influenza virus (AIV) surveillance in the LBMs and to develop new and innovative methods to prevent the introduction of AIV into the LBMs and to find ways to eliminate it when it is detected.

Live-bird markets (LBMs) have been intensely studied in recent years because avian influenza viruses in the markets are closely associated with avian influenza in commercial poultry and the markets may serve as a 'fertile ground' for virus mutations and emergence of new influenza viruses with increased virulence or ability to infect other species, including humans.

In 1997, an H5N1 avian influenza virus (AIV) emerged in Hong Kong LBMs to infect 18 people; 6 of whom died (Claas et al. 1998). The source of human infections was due to direct contact with infected chickens in the LBMs; there was no human-to-human spread. Subsequent studies on the H5N1 virus showed that the virus most likely evolved by the reassortment of virus genes from at least 3 different avian influenza viruses that were circulating within the LBMs in Hong Kong (Guan et al. 1999; Hoffmann et al. 2000).

In the United States, the LBMs were first recognized as a potential source of AIV in 1986 following the re-emergence in Pennsylvania of an H5N2 AIV of low pathogenicity believed to be the precursor virus that caused the outbreak of highly pathogenic H5N2 in 1983-84. The source of the low-pathogenic H5 virus was traced to the LBMs in the Northeast United States. Eradication of the H5N2 virus from the

[#] National Veterinary Services Laboratories, Veterinary Services, Animal and Plant Health Inspection Service, US Department of Agriculture, P. O. Box 844, Ames, Iowa 50010, USA. E-mail: dennis.senne@aphis.usda.gov

R. S. Schrijver and G. Koch (eds.), Avian Influenza, 19–24.
© 2005 *Springer. Printed in the Netherlands.*

LBMs was accomplished by the end of 1987. Since that time, extensive surveillance was conducted to monitor for AIV circulating in the LBMs. In 1994, an H7N2 AIV of low pathogenicity was detected in the LBMs and has persisted since, despite efforts to eliminate the virus. Since 1996, the LBMs have been implicated as a source of H7N2 AIV in at least 5 outbreaks in commercial poultry in the Northeast United States (Akey 2003; Davison et al. 2003; Dunn et al. 2003, D. Senne unpublished observation).

In this paper we: 1. briefly describe the LBM system in the USA; 2. review surveillance activities in the LBMs; 3. review the recent outbreaks of AI for which the LBMs were implicated as a source of virus in commercial poultry; 4. summarize the molecular changes in an H7N2 AIV that has continued to circulate in the markets since 1994; and 5. present past and future plans for the control of AIV in the LBMs.

Keywords: avian influenza; live-bird markets; surveillance; United States

What are live-bird markets?

The LBMs in the Northeast United States are part of a complex marketing system that provide a source of fresh poultry meat preferred by ethnic populations in many of the large cities. The northeastern LBM system is comprised of more than 120 markets in 6 states (New York, New Jersey, Connecticut, Rhode Island, Pennsylvania and Massachusetts) with the majority of the markets being located in New York and New Jersey. Birds entering the LBM system come from a variety of sources including farms that raise birds specifically for the LBMs, backyard flocks and commercial poultry farms. Most of the birds come from adjacent states but some birds are transported several hundred miles from states as far west as Ohio. Birds are collected at the farms by dealers and/or wholesalers and delivered by truck or vans to distribution centres within the city or directly to the LBMs where birds are placed in open holding pens or in cages. Cages are generally stacked 4 to 5 tiers high with separate food and water sources for each tier of cages. Customers can hand-pick birds they wish to purchase and the birds are then individually processed and the carcasses prepared according to the customers' specifications.

The LBMs provide an environment where reservoir species of the AIV, i.e. ducks and geese, are housed closely with chickens, turkeys, guinea fowl and quail etc., which are not natural hosts for the virus. Commingling of different avian species and daily introduction of new birds provides opportunity for the AIV to replicate and adapt to new hosts and the infection to persist within the market system for extended periods. Long-term replication of AIV in unnatural host may facilitate accumulation of point mutations which could lead to increased virulence.

Surveillance activities

Each year since 1994, between 1,457 and 8,120 tracheal- and cloacal-swab pools were collected from the LBMs in the Northeast United States and tested for presence of AIV by virus isolation in embryonated chicken eggs (Table 1). In 1994, an H7N2 AIV of low pathogenicity was introduced into the LBMs which continued to circulate in the market system. During 1994-2003, between 30 and 808 isolations of H7N2 AIV were made each year (Table 1). In addition to the H7N2 virus, several introductions of H5 and other H7 subtypes were detected but the latter subtypes did not become established. The sporadic detection of the H5 and H7N3 subtype viruses suggests that these viruses are not well adapted to poultry and disappear when the infected birds are sold and removed from the markets.

Table 1. Number of samples tested and number of isolations of avian influenza virus subtypes H5 and H7 from live-bird markets of the Northeast United States, FY 1994-2003

Fiscal year	Total no. tested	No. isolates (of H5 subtypes*)	No. isolates (of the H7N2 subtype*)	No. isolates (of H7* subtypes)
1994	1,791		30	1 (H7N3)
1995	5,214		170	
1996	1,740		363	
1997	2,060		188	
1998	2,497		372	
1999	3,679	3 (H5N2)	808	10 (H7N3)
2000	1,457		185	
2001	2,756	5 (H5N2)	419	
2002	8,120	3 (H5N2)	745	
2003	5,709	4 (H5N8) 1 (H5N9)	354	

* All H5 and selected H7 avian influenza viruses were characterized as low-pathogenicity viruses

Outbreaks in commercial poultry linked to the LBMs

Since 1996, five outbreaks of low-pathogenic H7N2 in commercial poultry have been linked to the LBMs in Northeastern United States as the source of infection: 1) Pennsylvania, in 1996-97 (18 layer, 2 layer pullet, and 1-meat turkey farms); 2) Pennsylvania, in 2001/2002 (5 broiler, and 2 broiler breeder farms); 3) Virginia/West Virginia/North Carolina in 2002 (210 flocks, 4.7 million birds involving turkey breeders, meat turkeys, broiler breeders, broilers, layers and quail); 4) Connecticut, in 2003 (4 layer farms involving 3.9 million birds); and 6) Rhode Island, in 2003 (32,000 layers). A more detailed account of the outbreaks is given by other authors on this volume (D. Swayne, D. Senne).

A direct epidemiologic connection between the LBMs and the commercial-poultry outbreaks noted above was established in only two outbreaks; the 1996-97 outbreak in Pennsylvania and the outbreak of 2003 in Rhode Island. In both outbreaks, trucks hauling birds to the LBMs had been on the affected premises within a week before the appearance of clinical disease. However, in the remainder of the outbreaks, the LBMs were implicated as the source of viruses because the causative virus was genetically indistinguishable from the H7N2 virus present in the LBMs, the only known source of this strain of AIV. In 2001, more than 185 farms that routinely supplied birds to the LBMs were surveyed for presence of AIV and specific antibodies to AIV in an attempt identify possible sources of the H7N2 virus outside of the LBM system. No H7N2 virus or specific antibodies were detected in the 2,225 swab specimes or 2,450 serums.

Molecular changes in the low-pathogenic H7N2 virus since 1994

The continued circulation of an H7N2 virus of low pathogenicity in the LBM system since 1994 has provided an opportunity to study the genetic changes in the virus following replication in unnatural hosts for an extended period, especially as it relates to the amino-acid motif at the cleavage site of the H protein. There are four reports where low-pathogenic H5 and H7 viruses, after circulating in poultry for 1 to 9 months, have mutated to highly pathogenic viruses (Banks et al. 2001; Horimoto et al. 1995; Kawaoka, Naeve and Webster 1984; Senne et al. 2002).

Since 1994, several changes in the amino-acid motif have been observed at the cleavage site of the H protein of the H7N2 virus in the LBMs suggesting that the virus is progressing toward becoming highly pathogenic. From 1994 to 2002, the cleavage-site motif gradually changed from PENPKTR/GLF to PEKPKKR/GLF, with the addition of two lysine (K) residues at the -2 and -5 positions of the HA1 (Table 2). In each case, where a virus with a new motif was detected, it eventually replaced the previous virus and became the predominant strain in the markets. For example, in 1994, the amino-acid sequence at the cleavage site of the H protein contained two basic amino acids and was PENPKTR/GLF. In 1996, 1 of 11 isolates tested had a motif containing three basic amino acids (PEKPKPR/GLF), with a K to N (asparagine) substitution at the -5 position. By 1998, the motif with three basic amino acids became predominant and was the only motif detected in isolates tested in 2000 and 2001. In 2002, a fourth basic amino acid was observed in isolates from several markets. This change came about by a P (proline) to K substitution at the -2 position, resulting in a cleavage-site motif of PEKPKKR/GLF. It is possible that the H7N2 virus with four basic amino acids may become highly pathogenic with the acquisition of an additional basic amino acid. The detection of an H7N2 virus with four basic amino acids provided increased urgency to implement a planned 3-day closure of all LMBs in the Northeast United States in an attempt to rid the markets of low pathogenic H7N2 AIV. Details of the market closures are described elsewhere in this chapter.

Table 2. Deduced amino-acid sequences of H7N2 avian influenza viruses isolated from live-bird markets of Northeast United States, FY 1994-03

Fiscal year	Amino-acid sequence at HA cleavage site*	No. isolates tested with sequence	Percent isolates with sequence
1994	PEN**PKTR**/GLF	1	100
1995	PEN**PKTR**/GLF	6	55
	PEN**PKPR**/GLF	4	36
	PEN**PKIR**/GLF	1	9
1996	PEN**PKPR**/GLF	10	91
	PE**KPKTR**/GLF	1	9
1997	PEN**PKPR**/GLF	3	100
1998	PEN**PKPR**/GLF	12	92
	PE**KPKPR**/GLF	1	8
1999	PEN**PKPR**/GLF	4	29
	PE**KPKPR**/GLF	10	71
2000	PE**KPKPR**/GLF	20	100
2001	PE**KPKPR**/GLF	22	100
2002	PE**KPKKR**/GLF	9	31
	PE**KPKPR**/GLF	20	69
	Market Closures (April 2002)		
2003	PE**KPKPR**/GLF	28	100

*Basic amino acids (lysine[K] and arginine [R]) are shown in bold print

Between 1994 and 2003, other molecular changes were detected in H7N2 viruses circulating in the markets. In 1995, a threonine (T) to P substitution at the -2 position was first observed. By 1997 this change was observed in most isolates. Proline at the -2 position is unique among H7 subtypes and has been used as a marker to distinguish LBM H7 viruses from other low-pathogenic H7 subtypes isolated outside the LBM system. In addition to the changes at the cleavage site of the H protein, a 24-nucleotide deletion, believed to be part of the receptor-binding region, was first observed in LBM H7N2 isolates in 1996 and is seen in all isolates since 2000. The significance of this deletion is not known but it could be related to the adaptation of the H7N2 virus to poultry.

Past, present and future control activities in the LBMs

In the United States, the authority to control outbreaks of low-pathogenic AIV is given to the individual states, whereas outbreaks of highly pathogenic AIV are managed at the federal level with assistance from the state authorities and the poultry industry. Therefore, control of the low-pathogenic avian influenza (LPAI) in the LBM system has been the responsibility of individual states.

Control of LPAI in the LBMs will differ from state to state; primary focus has been on education, increased surveillance and sanitation, and the adoption of better laws to give states more authority to deal with LPAI. Most states require AIV-positive LBMs to sell off the bird inventory, thoroughly clean and disinfect the premises and leave the market empty of birds from 1 to 3 days before repopulating the market. This approach was successfully used in 1986 and 1993 when low-pathogenic H5N2 AIV was found in LBMs in Northeastern United States, but it has not worked for the H7N2 virus currently present in the LBM system. Surveillance conducted in 2001 showed that up to 60% of the LBMs were positive for the low-pathogenic H7H2 virus that has persisted in the LBMs in Northeast United States since 1994.

The lack of success in reducing the number of H7N2-positive markets prompted the US Department of Agriculture (in March of 1999) to establish a LBM working group. The working group was comprised of representatives from state and federal governments as well as representatives from the LBMs and commercial poultry industry. The charge of the working group was to develop plans to control LPAI in the LBMs. One of the recommendations from the group was to implement a system-wide closure of the retail LBMs in the Northeast United States. The market closure was implemented April 8-10, 2002. Details of the market closures are described elsewhere (Mullaney 2003). In summary, birds in all markets were sold or killed and the markets were thoroughly cleaned and disinfected. Each market was then inspected to ensure that the cleaning and disinfection were properly completed and environmental samples were collected 24 hrs post-cleaning and tested for presence of AIV. The markets were required to stay empty of all birds for a 3-day period. Market owners were paid $3,000 each to cover lost revenue during the closure period. Funding for the closure (>$900,000 USD) was provided by Animal and Plant Health Inspection Service (APHIS) contingency funds.

Following the closure, markets were repopulated with birds only from AIV-monitored (AIV-negative) source flocks. Surveillance showed that the LBMs remained negative for AIV for about 5 weeks before the H7N2 virus was again detected. It is not known if the virus persisted in the markets or was reintroduced into the markets. However, since the closure the H7N2 virus with four basic amino acids has not been detected.

The control of AIV in LBMs has been identified as a critical component of the proposed LPAI control programme in the United States. The LPAI control programme will most likely be a joint state and federal programme with costs for monitoring and flock indemnities being shared by US Department of Agriculture and participating states. The programme will also consider new approaches to aid in the control of LPAI in LBMs. These may include the licensing of dealers/wholesalers to ensure that birds are obtained only from monitored flocks and that proper cleaning and disinfection of trucks and equipment are performed before going to the farms to pick up birds. Also, vaccination to reduce the susceptibility of the birds in the markets and individual bird identification to trace sources of virus outside the LBM system are being considered.

The LBMs are a reservoir of AIV for commercial poultry and can provide a favourable environment for the emergence of viruses with adaptive changes that can alter host specificity and/or virulence. Therefore, efforts must continue to find ways to prevent the introduction of AIV into the markets and to eliminate AIV when it is introduced

References

Akey, B.L., 2003. Low-pathogenicity H7N2 avian influenza outbreak in Virginia during 2002. *Avian Diseases,* 47 (special issue), 1099-1103.

Banks, J., Speidel, E.S., Moore, E., et al., 2001. Changes in the haemagglutinin and the neuraminidase genes prior to the emergence of highly pathogenic H7N1 avian influenza viruses in Italy. *Archives of Virology,* 146 (5), 963-973.

Claas, E.C.J., Osterhaus, A.D.M.E., Van Beek, R., et al., 1998. Human influenza A H5N1 virus related to a highly pathogenic avian influenza virus. *Lancet,* 351 (9101), 472-477.

Davison, S., Eckroade, R.J., Ziegler, A.F., et al., 2003. A review of the 1996-98 nonpathogenic H7N2 avian influenza outbreak in Pennsylvania. *Avian Diseases,* 47 (special issue), 823-827.

Dunn, P.A., Wallner-Pendleton, E.A., Lu, H., et al., 2003. Summary of the 2001-02 Pennsylvania H7N2 low pathogenicity avian influenza outbreak in meat type chickens. *Avian Diseases,* 47 (special issue), 812-816.

Guan, Y., Shortridge, K.F., Krauss, S., et al., 1999. Molecular characterization of H9N2 influenza viruses: were they the donors of the "internal" genes of H5N1 viruses in Hong Kong? *Proceedings of the National Academy of Sciences of the United States of America,* 96 (16), 9363-9367.

Hoffmann, E., Stech, J., Leneva, I., et al., 2000. Characterization of the influenza A virus gene pool in avian species in Southern China: was H6N1 a derivative or a precursor of H5N1? *Journal of Virology,* 74 (14), 6309-6315.

Horimoto, T., Rivera, E., Pearson, J., et al., 1995. Origin and molecular changes associated with emergence of a highly pathogenic H5N2 influenza virus in Mexico. *Virology,* 213 (1), 223-230.

Kawaoka, Y., Naeve, C.W. and Webster, R.G., 1984. Is virulence of H5N2 influenza viruses in chickens associated with loss of carbohydrate from the hemagglutinin? *Virology,* 139 (2), 303-316.

Mullaney, R., 2003. Live-bird market closure activities in the Northeastern United States. *Avian Diseases,* 47 (special issue), 1096-1098.

Senne, D.A., Pedersen, J.C., Mathieu, C., et al., 2002. Characterization of a novel highly pathogenic avian influenza virus from Chile. *In: Proceedings of the American Veterinary Medical Association, July 13-17, 2002, Nashville, TN.*

4

Influenza A virus surveillance in wild birds

V. Munster[#], A. Wallensten[##], B. Olsen[###], G.F. Rimmelzwaan[#], A.D.M.E. Osterhaus[#] and R.A.M. Fouchier[#,*]

Abstract

Surveillance studies in wild animals provide information on the prevalence of avian influenza viruses in the environment, and enables banking of reference reagents and putative vaccine strains to be used in times of outbreaks in humans and animals. In the past five years we have performed surveillance studies in wild birds primarily in The Netherlands and Sweden. In ducks, geese and gulls the prevalence of influenza A virus ranged from 0 to 60 percent, depending on bird species, location and season. The detection of avian influenza A viruses in other bird species was rare. Most of the fifteen haemagglutinin subtypes and all nine neuraminidase subtypes described to date were found in birds in Northern Europe. In addition, we have identified a novel haemagglutinin subtype(H16) in black-headed gulls. Viruses of subtypes H5 and H7 were found less frequently than other subtypes, and were closely related to the H5 and H7 highly pathogenic avian influenza viruses that have caused outbreaks in poultry in Italy and The Netherlands between 1997 and 2003.
Keywords: bird; poultry; avian influenza; *Orthomyxovirus*; surveillance; ecology; prevalence; PCR; subtype

Introduction

Influenza virus types A, B and C all belong to the family of *Orthomyxoviridae* and have therefore many biological properties in common (Murphy and Webster 1996). A key difference between them is their in-vivo host range; whereas influenza viruses of types B and C are predominantly human pathogens that have also been isolated from seals and pigs, respectively (Osterhaus et al. 2000; Guo et al. 1983), influenza A viruses have been isolated from many species including humans, pigs, horses, marine mammals and a wide range of domestic and wild birds (Webster et al. 1992). It is generally accepted that in the human influenza pandemics from the last centuries and numerous outbreaks in domestic and wild animals, interspecies transmission of avian influenza viruses has played a crucial role (Webster et al. 1992).

Predominantly water-associated wild birds such as ducks, geese, gulls and shorebirds form the reservoir of influenza A viruses in nature (Figure 1). All fifteen antigenic subtypes of the virus surface glycoprotein haemagglutinin and all nine

[#] National Influenza Center and Department of Virology, Erasmus MC, Rotterdam, The Netherlands
[##] Smedby Health Care Centre, Kalmar, Sweden
[###] Department of Infectious Diseases, Umeå University, SE-901 87 Umeå, Sweden.
[*] Corresponding author: Dr. R.A.M. Fouchier, Dept. Virology, Erasmus MC, PO Box 1738, 3000 DR Rotterdam, The Netherlands. E-mail: r.fouchier@erasmusmc.nl

R. S. Schrijver and G. Koch (eds.), Avian Influenza, 25–30.
© 2005 *Springer. Printed in the Netherlands.*

subtypes of neuraminidase that have been identified to date have been isolated from these bird species (Webster et al. 1992). Avian influenza viruses preferentially infect cells lining the intestinal tract of birds and are excreted in high concentrations in their faeces. The transmission of influenza viruses between birds is thought to occur primarily via the faecal-oral route. Whereas avian influenza viruses are generally non-pathogenic in their natural hosts, they may cause significant morbidity and mortality upon transmission to other species.

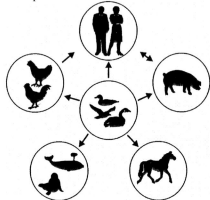

Figure 1. The natural hosts of influenza A virus. Wild aquatic birds are thought to form the influenza A virus reservoir in nature, from which viruses can be transmitted to poultry, pigs, horses, marine mammals and humans. Influenza A viruses can also be transmitted from poultry and pigs to humans

It is virtually impossible to prevent outbreaks of influenza A virus infection in domestic animals and to prevent novel influenza pandemics in man. Through the continuous surveillance of wild birds around the world the prevalence of influenza A viruses in the environment can be monitored and the pathogenic and antigenic properties of the circulating viruses can be determined. In addition, such studies will ensure that panels of reference reagents required for testing of animals and man can be updated continuously and that virus isolates with the appropriate antigenic properties will be available for the generation of potential influenza vaccines (Fouchier et al. 2003).

Previously, we have described a RT-PCR-based screening procedure for influenza A virus that is suitable for the detection of all influenza A virus strains identified to date (Fouchier et al. 2000). This screening method is more sensitive than classical virus propagation in eggs and cell cultures and provides a good alternative for rapid high-throughput screening for influenza A viruses in wild birds. Using this method and with the help of a large network of ornithologists, we have screened a wide spectrum of wild birds of different taxa in Northern Europe since 1998 for the presence of influenza A virus. Although our primary focus was on ducks, geese and gulls because these species are known to be susceptible to influenza A virus, we have also tested thousands of samples from passerines, waders, birds of prey, etc.

Materials and methods

Specimens
Birds were trapped using duck traps, wader funnel traps, mist nets, clap nets or Helgoland funnel traps. For bird species that could not be trapped, fresh dropping

samples were collected from the ground at locations where large numbers of birds congregate. Cloacal swabs and fresh dropping samples were collected using cotton swabs and subsequently stored in transport media at -70°C. From small birds, mainly passerines and small shorebirds, we collected fresh dropping samples rather than cloacal swabs after capture because the cotton swabs were too large. Transport media consisted of Hanks balanced salt solution supplemented with 10 % glycerol, 200 U/ml penicillin, 200 µg/ml streptomycin, 100 U/ml polymyxin B sulphate, 250 µg/ml gentamicin and 50 U/ml nystatin (all from ICN, Zoetermeer, The Netherlands). Most duck and goose samples were collected in the autumn and winter whereas for most passerine birds and shorebirds samples were collected in spring, summer and early autumn.

RNA isolation

RNA was isolated using a high-pure RNA isolation kit (Roche Molecular Biochemicals) according to the instructions from the manufacturer, with minor modifications. A 0.2 ml sample was homogenized by vortexing and subsequently lysed with 0.4 ml lysis/binding buffer. After binding to the column, DNase-I digestion and washing, the RNA was eluted in 50 µl nuclease-free double-distilled water. Initially, pools of 5 samples are tested (without significant loss of sensitivity). Upon identification of positive pools, the individual positive samples were identified.

RT-PCR

Samples were amplified in a one-step RT-PCR in 25 µl final volume, containing 50 mM Tris.HCl pH 8.5, 50 mM NaCl, 7 mM $MgCl_2$, 2 mM DTT, 1 mM each dNTP, 0.4 µM each oligonucleotide, 2.5 U recombinant RNAsin, 10 U AMV reverse transcriptase, 2.5 U Ampli-Taq DNA polymerase (all enzymes from Promega Benelux BV, Leiden, The Netherlands) and 5 µl RNA. Primers M52C (5'- CTT CTA ACC GAG GTC GAA ACG -3') and M253R M253R (5'- AGG GCA TTT TGG ACA AAG/T CGT CTA -3') were used. Thermo-cycling was performed in an MJ PTC-200 apparatus using the following cycling conditions: 30 min. at 42°C, 4 min. at 95°C once; and 1 min. at 95°C, 1 min. at 45°C, 3 min. at 72°C repeated 40 times. Each reaction (5 µl/sample) was analysed by dot-blot hybridization.

Dot-blot hybridization

Five µl of the RT-PCR products was incubated for 5 min. at room temperature with 45 µl 10 mM Tris.HCl pH 8.0, 1 mM EDTA and 50 µl 1 M NaOH for denaturation. Samples were transferred to prewetted Hybond N+ membranes (Amersham Pharmacia Biotech Benelux, Roosendaal, The Netherlands) using a dot-blot apparatus while applying vacuum. Samples were then treated for 3 min. with 0.1 ml 1 M Tris.HCl pH8.0, after which vacuum was applied again for 10 sec, and the membrane removed from the apparatus. Blots were washed three times for 10 min. with 0.3 M NaCl, 30 mM Na-citrate pH 7, dried, and stored at 4°C. Blots were prehybridized for 5 min. at 55°C in 2 x SSPE (0.3 M NaCl, 20 mM NaH_2PO_4, 2 mM EDTA, pH 7.4) and 0.1 % SDS, after which biotinylated oligonucleotide probe Bio-M93C (5'- CCG TCA GGC CCC CTC AAA GCC GA –3') was added to 2 pmol/ml and hybridization was continued for 45 min. at 55°C. Blots were washed twice for 10 min. at 55°C with hybridization buffer, transferred to 2 x SSPE with 0.5 % SDS after which streptavidin peroxidase (Roche Molecular Biochemicals) was added to 0.125 U/ml and incubated for 45 min. at 42°C. Blots were washed 10 min. at 42°C in 2 x SSPE, 0.5 %SDS, 10 min. at 42°C in 2 x SSPE, 0.1 % SDS and 10 min. at room temperature in 2 x SSPE, after which the samples were visualized using ECL detection reagents and exposure to hyperfilm (Amersham Pharmacia Biotech Benelux, Roosendaal, The Netherlands) for 5 - 60 seconds.

Virus isolation and characterization

For all RT-PCR-positive samples, 200 µl of the original specimen was inoculated in the allantoic cavities of 11-day-old embryonated chicken eggs (Rimmelzwaan et al. 1998). When necessary, the allantoic fluid was harvested and passaged again in embryonated eggs. HA titers in virus stocks were determined with turkey erythrocytes using standard procedures. Virus isolates were characterized by haemagglutination and neuraminidase-inhibition assays with subtype-specific hyper-immune rabbit antisera raised against HA/NA preparations of virus isolates representing all 15 HA subtypes (Rohm et al. 1996). Sequencing was also conducted of HA and NA genes.

Results

PCR-based detection of influenza A virus

From 1998 to 2003 we have collected cloacal swabs and fresh dropping samples from more than 15,000 birds that were tested for the presence of influenza A virus RNA by RT-PCR. The majority of these samples were collected at different locations in The Netherlands (Fouchier et al. 2004) and Sweden (Wallensten et al. 2003). Samples were collected from 252 different bird species with the majority of samples originating from ducks, geese, gulls and shorebirds. RT-PCR positive samples were obtained only from geese (white-fronted, greylag and brent geese), ducks (mallard, wigeon, shoveler and teal), black-headed gulls and guillemots (Table 1). For ducks, gulls and guillemots almost exclusively cloacal swabs were tested, whereas geese samples included both cloacal swabs and fresh droppings, both of which included influenza A virus-positive samples. With 345 RT-PCR-positive samples, the overall detection of influenza A virus was approximately 2.3 percent of all samples.

Table 1. RT-PCR-based screening of wild birds for the presence of influenza A virus

Order (No. species tested)	No. tested	No. PCR-positive (%)	PCR-positive bird species
Anseriformes (22)			
Geese (8)	1627	20 (1.2)	White-fronted, Greylag, Brent goose
Ducks (12)	5223	301 (5.8)	Wigeon, Mallard, Shoveler, Teal
Others (2)	93	0	
Charadriiformes (39)			
Gulls (5)	1606	10 (0.6)	Black-headed gull
Guillemots (2)	30	3 (10.0)	Guillemot
Others (32)	2130	0	
Others (181)	4348	0	

However, at selected locations the prevalence in specific species was much higher at some occasions. For instance, in October 1999 14 of 132 (10.6 %) of all mallard ducks in duck trap 'Bakkerswaal' in Lekkerkerk, The Netherlands were positive for influenza A virus and in the second week of August 1999 6 out of 10 (60 %) black-headed gulls in Ottenby, Sweden were positive.

Virus isolation and characterization

Virus could be isolated in embryonated chicken eggs for about fifty percent of the 345 RT-PCR-positive cloacal swabs and dropping samples. Most of the samples that remained negative even upon repeated attempts to isolate the virus contained relatively low copy numbers of viral RNA as judged by the low intensity of the bands upon hybridization of dot-blots. A low proportion of RT-PCR-positive samples that appeared to contain relatively high copy numbers of viral RNA may have been stored improperly in the field, or may contain virus that cannot be isolated efficiently in embryonated chicken eggs.

Most of the fifteen haemagglutinin subtypes (except subtypes 9, 14, 15) and all nine neuraminidase subtypes described throughout the world to date were found in birds in Northern Europe in the past five years. The haemagglutinin gene of 4 virus isolates obtained from black-headed gulls could not be identified using our panels of reference reagents; these HA genes represent a novel subtype, H16.

Viruses of subtypes H5 and H7 were found less frequently than other subtypes. Sequence analyses of haemagglutinin and neuraminidase genes of influenza A viruses obtained from mallard ducks indicate that these viruses are very closely related to the H5 and H7 highly pathogenic avian influenza viruses that have caused outbreaks in poultry in Italy and The Netherlands between 1997 and 2003.

Discussion

This study demonstrates that influenza A viruses are highly prevalent in wild birds in Northern Europe. For ducks (~5200 tested), geese (~1600 tested) and guillemots, the prevalence varied from 0 to 11 %, and for gulls (~1600 tested) from 0 to 60 % dependent on the birds' age, time, location and species. Using a highly sensitive RT-PCR procedure, we only found evidence for influenza A virus infection of aquatic birds (ducks, geese, gulls, guillemots), which is in agreement with previous studies performed by many laboratories using virus isolation in embryonated chicken eggs (Webster et al. 1992; Ito and Kawaoka 1998), but not in more than 6500 samples from other bird species (252 bird species were included in our dataset). A clear difference with other studies (Ito and Kawaoka 1998) is the apparent absence of influenza A virus in our collection of shorebirds. In Northern Europe, the prevalence of avian influenza in ducks and geese is at its peak between late summer and early winter when the birds leave their breeding grounds and start migrating, and up to 30 % of a flock or colony may be excreting virus

Many different influenza A virus subtypes were found to circulate at the same time, in the same bird species at a single location in The Netherlands. For instance, in duck trap 'Bakkerswaal' in Lekkerkerk, The Netherlands, 14 out of 132 mallard ducks caught in October 1999 were positive for influenza A virus, and HA subtypes 1, 2, 4, 5 and 11 were identified. The genetic and antigenic heterogeneity observed for some of our virus isolates indicates that continuous influenza A virus surveillance is required to keep panels of reference reagents and potential future vaccine strains updated.

Viruses of subtypes H5 and H7 were found less frequently than other subtypes. Sequence analyses of haemagglutinin and neuraminidase genes of influenza A viruses obtained from mallard ducks indicate that these viruses are very closely related to the H5 and H7 highly pathogenic avian influenza viruses that have caused outbreaks in poultry in Italy and The Netherlands between 1997 and 2003 (Capua and Marangon 2000; Banks et al. 2000; Fouchier et al. 2004).

Acknowledgments

We are grateful to the numerous ornithologists that have made this work possible. This work was sponsored in part by grants from the Netherlands Ministry of Agriculture, the European Union and the Research Council of Southeastern Sweden (FORSS). R.F. is a fellow of the Royal Netherlands Academy of Arts and Sciences.

References

Banks, J., Speidel, E.C., McCauley, J.W., et al., 2000. Phylogenetic analysis of H7 haemagglutinin subtype influenza A viruses. *Archives of Virology,* 145 (5), 1047-1058.

Capua, I. and Marangon, S., 2000. The avian influenza epidemic in Italy, 1999-2000: a review. *Avian Pathology,* 29 (4), 289-294.

Fouchier, R.A.M., Bestebroer, T.M., Herfst, S., et al., 2000. Detection of influenza A viruses from different species by PCR amplification of conserved sequences in the matrix gene. *Journal of Clinical Microbiology,* 38 (11), 4096-4101.

Fouchier, R.A.M., Olsen, B., Bestebroer, T.M., et al., 2003. Influenza A virus surveillance in wild birds in Northern Europe in 1999 and 2000. *Avian Diseases,* 47 (special issue), 857-860.

Fouchier, R.A.M., Osterhaus, A.D.M.E. and Brown, I.H., 2003. Animal influenza virus surveillance. *Vaccine,* 21 (16), 1754-1757.

Fouchier, R.A.M., Schneeberger, P.M., Rozendaal, F.W., et al., 2004. Avian influenza A virus (H7N7) associated with human conjunctivitis and a fatal case of acute respiratory distress syndrome. *Proceedings of the National Academy of Sciences of the United States of America,* 101 (5), 1356-1361.

Guo, Y.J., Jin, F.G., Wang, P., et al., 1983. Isolation of influenza C virus from pigs and experimental infection of pigs with influenza C virus. *Journal of General Virology,* 64 (1), 177-182.

Ito, T. and Kawaoka, Y., 1998. Avian influenza. *In:* Hay, A.J. ed. *Textbook of influenza.* Blackwell, Oxford, 126-136.

Murphy, B.R. and Webster, R.G., 1996. Orthomyxoviruses. *In:* Howley, P.M. ed. *Fields virology.* 3rd edn. Lippincott-Raven, Philadelphia, 1397-1445.

Osterhaus, A.D.M.E., Rimmelzwaan, G.F., Martina, B.E.E., et al., 2000. Influenza B virus in seals. *Science,* 288 (5468), 1051-1053.

Rimmelzwaan, G.F., Baars, M., Claas, E.C.J., et al., 1998. Comparison of RNA hybridization, hemagglutination assay, titration of infectious virus and immunofluorescence as methods for monitoring influenza virus replication in vitro. *Journal of Virological Methods,* 74 (1), 57-66.

Rohm, C., Zhou, N.A., Suss, J.C., et al., 1996. Characterization of a novel influenza hemagglutinin, H15: criteria for determination of influenza a subtypes. *Virology,* 217 (2), 508-516.

Wallensten, A., Munster, V.J., Fouchier, R.A.M., et al., 2003. Avian influenza A virus in ducks migrating through Sweden. *In:* Kawaoka, Y. ed. *Options for the control of influenza V: proceedings of the international conference for the control of influenza V, Okinawa, Japan, October 7-11, 2003.* Elsevier, Amsterdam, 771-772. International Congress Series no. 1263.

Webster, R.G., Bean, W.J., Gorman, O.T., et al., 1992. Evolution and ecology of influenza A viruses. *Microbiology Reviews,* 56 (1), 152-179.

OUTBREAKS IN DENSELY POPULATED POULTRY AREAS

5

The control of avian influenza in areas at risk: the Italian experience 1997-2003

S. Marangon[#], I. Capua[#], E. Rossi[#], N. Ferre'[#], M. Dalla Pozza[#], L. Bonfanti[##], A. Mannelli[###]

Abstract

From 1997 to 2003, Italy has been affected by two epidemics of highly pathogenic avian influenza and by several outbreaks of low-pathogenic avian influenza (LPAI). In 1999-2000 a severe avian influenza (AI) epidemic affected the country. The epidemic was caused by a type-A influenza virus of the H7N1 subtype, originated from a low-pathogenic (LP) AI virus, which spread in 1999 among poultry farms in Northeastern Italy and eventually became virulent by mutation with the emergence of a highly pathogenic (HP) strain. From 17 December 1999 to 5 April 2000, a total of 413 outbreaks (178 meat-type turkey, 5 turkey breeder, 29 broiler breeder, 119 layer, 37 broiler, 9 guinea fowl, 11 game farm and 25 back-yard flocks) were identified and the last affected flock was stamped out on 5 April 2000. A total of about 16 million birds died or were stamped out on affected and at-risk premises.

From August 2000 to March 2001 in two epidemic waves, the H7N1 LPAI strain infected 73 turkey farms, which housed 1 million turkeys, 4 quail farms, with about 800,000 quails, and 1 layer farm (40,000 layers) located in the southwestern part of the Veneto Region (Verona and Padua provinces). To supplement disease-control measures already in force an emergency vaccination programme was implemented based on the 'DIVA' (differentiating infected from vaccinated animals) strategy. After the implementation of the vaccination programme, only 3 meat-type turkey farms were infected inside the vaccination area and among these, only one vaccinated flock was affected. The last affected flock was stamped out on 26 March 2001.

In October 2002, another LPAI virus of the H7N3 subtype emerged in the northern part of the country. The H7N3 LPAI strain rapidly spread among poultry flocks located in the densely populated poultry area (DPPA) which had been affected by the H7N1 epidemic in 1999-2001. Eradication measures were based on stamping out or controlled marketing of slaughterbirds on infected farms and on the prohibition of restocking of poultry farms. Restriction measures on the movement of live poultry, vehicles and staff were also imposed in the areas at risk. To supplement disease-control measures already in force, an emergency vaccination programme, based once again on the 'DIVA' strategy was drawn, approved by the EU Commission and

[#] Istituto Zooprofilattico delle Venezie, Viale dell'Università 10, 35020 Legnaro, Padova, Italy. E-mail: stefano.marangon@regione.veneto.it
[##] Unità Organizzativa Veterinaria, Direzione Generale Sanità, Regione Lombardia, Via Pola 11, 20100 Milano, Italy
[###] Dipartimento di Produzioni Animali, Epidemiologia ed Ecologia, Università di Torino, Via Leonardo da Vinci, 44, 10095, Grugliasco, Torino, Italy

R. S. Schrijver and G. Koch (eds.), Avian Influenza, 33–39.
© 2005 *Springer. Printed in the Netherlands.*

implemented in the area. From 10 October 2002 to 10 October 2003, the H7N3 LPAI virus was able to spread and infect a total of 387 poultry holdings, mainly meat-type turkey farms; of these 88 were vaccinated. The implementation of a vaccination programme and the enforcement of strict restriction measures did not avoid the spread of the H7N3 LPAI virus infection among meat-turkey farms located in a DPPA. Nevertheless, it was possible to prevent the massive spread of infection to poultry farms other than turkey and to neighbouring vaccinated areas.

Keywords: avian influenza; LPAI and HPAI; outbreak; Italy

Introduction

Since 1997 Italy has been affected by six epidemics of avian influenza caused by viruses of the H5 or H7 subtypes. These epidemics were caused by H5N2 HPAI, H7N1 (HPAI and LPAI) and H7N3 (LPAI) (Capua and Alexander 2004). The characteristics of these epidemics and their different impact on the poultry industry and possible control strategies are presented.

The most severe episode was observed during 1999-2001, in which Italy was affected by four subsequent epidemic waves of avian influenza caused by viruses of the H7N1 subtype. The first epidemic wave was caused by a low-pathogenicity avian influenza virus of the H7N1 subtype that subsequently mutated into a highly pathogenic avian influenza virus, after circulating in the industrial poultry population for approximately nine months. Following the emergence of the HPAI virus, which caused death or culling of about 16,000,000 birds on infected and at-risk farms, and the implementation of the measures indicated in Council Directive 92/40/CE, the LPAI virus re-emerged twice.

Following the eradication of the H7N1 virus an H7N3 virus was introduced in the industrial poultry population of Northern Italy. A vaccination programme based on the 'DIVA' strategy was implemented and is currently being used.

The present paper reports of the strengths and weaknesses of the control strategies implemented to deal with AI epidemics occurring in diverse conditions.

Materials and methods

HPAI eradication measures

The measures required under the European Union (EU) legislation (CEC 1992) to control the disease were enforced. The identification of the AI infected farms was based on the notification of suspected cases by farmers and on official inspections of flocks at risk of infection. Epidemiological investigations were carried out on the affected farms by means of a standardized questionnaire. HPAI-virus-infected and at-risk flocks were stamped out. In the areas at risk, the prohibition of restocking of poultry farms associated with the enforcement of restriction measures on the movement of live poultry, vehicles and staff was imposed.

LPAI (H7N1-H7N3) control and eradication measures

No compulsory eradication measures to control LPAI-virus infections are provided for in the current EU legislation. In order to avoid the spread of the LPAI virus and the possible re-emergence of a HPAI-virus strain, the Italian authorities put in force a series of control measures in affected and at-risk areas. These included: monitoring of flocks at risk of infection, stamping out or controlled marketing of slaughterbirds on infected farms, restriction policies to restocking and movement of live birds, vehicles and staff. Furthermore, in order to supplement these control measures, a 'DIVA' vaccination programme against the disease was implemented. This vaccination strategy was based on heterologous vaccination. Briefly, since the antigen that induces the production of neutralizing antibodies is represented by the haemagglutinin, a vaccine containing an isolate with a homologous H group and heterologous N, from the field strain, was used as a 'natural marker' vaccine. Vaccination was carried out on meat-type turkey farms, layer flocks, capon and cockerel farms. Other species and production types were not vaccinated. Serological testing of sentinel birds and a discriminatory test able to distinguish the different type of anti-neuraminidase antibodies (anti-N1 and anti-N3) were applied in vaccinated flocks to monitor the epidemiological situation.

Laboratory diagnosis

During the LPAI and HPAI epidemics the diagnostic guidelines reported in the EU legislation on AI were followed (CEC 1992). With reference to the LPAI epidemic, in case of HI-positive results in vaccinated flocks, a discriminatory test (Capua et al. 2003) was used to differentiate between vaccinated/non-exposed and vaccinated/field-exposed birds/farms.

Data analysis

Epidemiological data collected during the HPAI epidemic on each affected flock were used to trace back the origin of the infection. The characteristics of HPAI outbreaks were analysed using χ^2 test and Mann-Whitney U test for categorical and ordinal variables, respectively. The risk of neighbourhood spread, and the association between inter-event distance and time were evaluated by multivariate logistic regression model (PROC LOGISTIC, SAS system 8.2). Inter-outbreak distance ≤ 500 metres was used as outcome, and quartiles of inter-outbreak times were used as predictors. Another predictor was included in the model, taking value equal to 1 when outbreak pairs belonged to the same avian species, 0 otherwise.

Results

1997-1998 H5N2 HPAI epidemic

The epidemic consisted of a total of 8 outbreaks (Capua et al. 1999), in backyard or semi-intensive flocks, located in the Veneto and Friuli Venezia Giulia regions. Although the origin of the epidemic was not established, the epidemiological investigation allowed identifying risk factors in the affected farms, primarily the marketing of infected birds, presence of mixed species and rearing in the open of birds. The disease was eradicated by the prompt implementation of Directive 92/40/EEC. A total of 7,731 birds were depopulated and no further isolations of the H5N2 virus have been made to date.

1999-2000 H7N1 HPAI epidemic

From 17 December 1999 to 5 April 2000 413 HPAI-infected poultry farms, mainly located in the Po Valley, were detected. A total of 13,731,253 birds were culled or died on the affected farms, and more than 2 million animals were pre-emptively slaughtered on 80 premises at risk of infection. The infection was detected more frequently in turkey ($\chi2= 118.37$, P<0.0001) and layer farms ($\chi2= 373.04$, P<0.0001), which accounted for 73% of the outbreaks, in larger flocks (z= 5.895, P<0.0001), in poultry farms located on the plain (altitude ≤ 300 m) ($\chi2=37.27$, P<0.0001). Tracing

35

exercises carried out on affected premises allowed the identification of the possible origin of the infection in 66.3% of the outbreaks. In particular, the origin of infection could be attributed to: movement of animals (1.0%), indirect contacts at the time of loading for slaughter of female turkeys (8.5%), neighbourhood spread (within 1 km radius) (26.2%), lorries for the transport of feedstuff, litter and carcasses (21.3%), and other indirect contacts (e.g. exchange of manpower, machinery, equipment) (9.4%). The logistic regression model for space–time association was significant (likelihood ratio $\chi2 = 53.5$, 4 df, P<0.001) and showed no evidence of poor fit (Hosmer and Lemeshow test: $\chi2 = 0.6$, 3 df, P = 0.90). Inter-event distance ≤500 m and inter-event time ≤10 days (first quartile) were strongly associated, indicating rapid viral transmission among contiguous farms (Table 1). Outbreak pairs involving the same avian species were more likely to be within 500 m from one another than outbreaks in different species.

Table 1. Results of the logistic regression model for space–time association

Inter-event time	OR estimate	95% Wald confidence limits	
quartile 1 vs 4	13.2	3.2	55.1
quartile 2 vs 4	3.6	0.75	17.3
quartile 3 vs 4	2.3	0.45	11.9
same species	1.3	1.2	1.7

2000-2001 H7N1 LPAI epidemic

From August to November 2000, the H7N1 LPAI strain infected 51 meat-type turkey farms, which housed 845,000 turkeys, and 1 quail farm, with a total of 429,000 quails, located in the southern part of the province of Verona. Another 3 quail farms, with a total of 405,000 quails, located in a contiguous province were also affected. The latter, were functionally linked to the farm situated in the province of Verona. To supplement disease-control measures already in force, an emergency vaccination programme against the disease was implemented in this area. Vaccination was implemented according to the 'DIVA' strategy based on the use of an inactivated heterologous vaccine (A/ck/PK/95-H7N3). Shortly after the beginning of the vaccination programme (December 2000 to March 2001), the H7N1 LPAI virus infected 3 meat-type turkey farms in the vaccination area and 20 poultry holdings (19 turkey farms and 1 layer farm) located in a contiguous unvaccinated area. Only one vaccinated flock was affected, and the virus did not spread from this to other vaccinated farms. The last H7N1 LPAI infected poultry flock was stamped out on 26 March 2001, and the results of the serological surveillance carried out both within and outside the vaccinated area to assess the possible presence of AI infection, demonstrated that the H7N1 AI virus strain was not circulating anymore. The application with negative results, of the 'DIVA' discriminatory test, was considered an additional guarantee for Member States, and on 30 November 2001, Commission Decision 2001/847/EC, authorized the marketing of fresh poultry meat obtained from vaccinated birds for intra-community trade.

2002-2003 H7N3 LPAI epidemic

In October 2002, another LPAI virus of the H7N3 subtype was introduced in the northern part of the country. The H7N3 LPAI strain rapidly spread among poultry flocks located in the densely populated poultry area (DPPA) that had been affected by the H7N1 epidemic in 1999-2001. The vaccination programme was based once again on a 'DIVA' strategy and was carried out using an AI inactivated heterologous vaccine (strain A/ck/IT/1999-H7N1). The beginning of the DIVA vaccination campaign was delayed up to 31 December 2002, due to unavailability of an appropriate vaccine. From 10 October 2002 to 10 October 2003, the H7N3 LPAI virus was able to spread and infect a total of 387 poultry holdings: 332 meat-type turkey, 5 turkey breeder, 12 broiler breeder, 13 layer, 6 guinea fowl, 4 broiler, 3 quail, 1 meat-duck farms and 11 back-yard flocks mainly located in the southern part of the two Italian regions. A total of 7,659,303 birds were involved in the epidemic, and among these 4,230,750 animals were stamped out in 163 affected flocks. The remaining 3,428,553 slaughterbirds were subjected to controlled marketing. Of the affected farms, 88 were vaccinated flocks. The first outbreak in a vaccinated flock occurred on 18 April. All the infected vaccinated flocks were meat turkeys mainly located in a limited area of the southern part of Verona province. It is interesting to point out that despite the poultry density in the latter area only 2 unvaccinated poultry farms (1 broiler breeder and 1 meat-duck farms) were affected. These farms were located in close proximity to previously vaccinated meat-turkey farms which had been field-exposed. Stamping-out measures or controlled marketing were enforced in all infected flocks, which housed a total of 1,523,320 birds.

The implementation of a vaccination programme and the enforcement of strict restriction measures did not avoid the spread of LPAI-virus infection among meat-turkey farms located in a DPPA. Nevertheless, it was possible to prevent the massive spread of infection to poultry farms not rearing turkeys and to neighbouring vaccinated areas. Only sporadic outbreaks of LPAI infection were detected in unvaccinated poultry farms, mainly located outside the vaccination area: 3 meat-turkey, 5 dealer and 2 quail farms in Lombardia, 1 dealer farm in Piemonte, and 2 meat-turkey farms in the Emilia-Romagna region. These flocks were promptly identified and stamped out.

Discussion

A few considerations can be made from retrospectively analysing the experience gained in the past 6 years with avian influenza in Italy. Firstly, Northeastern Italy can definitely be considered an area 'at risk' for avian influenza infections. This is also supported by AI epidemics which have occurred in the past (Franciosi et al. 1981; Petek 1982; Meulemans 1986; Papparella, Fioretti and Menna 1994; 1995) caused by viruses of the H6 and H9 subtypes. This could probably be linked both to the great numbers of wild birds which fly over the area during their migration and to the great numbers of imports of live birds into the area. For this reason, and considering the poultry density in the area, it is imperative that surveillance programmes are implemented to diagnose AI infections promptly.

The comparison between the 1997-1998 and 1999-2000 epidemics points out that if HPAI is diagnosed promptly and is not preceded by extensive circulation of the LPAI progenitor, the application of the measures imposed by Directive 92/40/EEC are efficacious in disease eradication. The devastating impact of the HPAI H7N1 epidemic in 1999-2000 was linked to loss of control of infection, primarily due to the previous circulation of the LPAI virus, which caused difficulties in identifying infected flocks promptly. Clearly, eradication efforts are more successful if there is no massive spread into the industrial circuits of intensively reared poultry.

The Italian 1999-2000 AI epidemic also emphasized that farmers and private companies should bear well in mind that within the current European legislation there is no financial aid from local or national governments or from the European Union in case of LPAI. Therefore, voluntary and permanent surveillance programmes should be implemented in order to allow the prompt diagnosis of infection by H5 and H7 LPAI viruses, to allow the enforcement of restriction and eradication policies until this is economically feasible.

The control of LPAI infections in DPPA is a challenging experience. A co-ordinated set of control measures including the application of adequate biosecurity measures, the enforcement of restriction policies to restocking and movement of live birds, vehicles and staff, and the implementation of a vaccination programme and of intensive monitoring measures in the areas at risk of infection, may have different outcomes on the basis of a series of variables. These include primarily the biological characteristics of the strain, the animal species and density at the moment of AI introduction and the functional organization of both the poultry industry and the veterinary services in the area. However, the availability of a well-structured legal basis for LPAI control, the prompt availability of vaccine, the general economic situation and the motivation of farmers and companies to eradicate the infection also play a major role in the eradication of avian influenza infections.

The experience gathered during the Italian 1997-2003 AI epidemics suggests that countries at risk of infection should have contingency plans and a general preparedness in order to deal appropriately with such infections. Outbreaks caused by avian influenza viruses of the H5 and H7 subtypes can no longer be considered rare events and therefore alternative strategies to a stamping-out policy should be considered, particularly for outbreaks occurring in densely populated poultry areas.

In our opinion it is imperative that this disease is dealt with as a problem of the industry and of veterinary public-health services. The different sets of data that are generated from surveillance and control programmes at the industry level must be made available to support decision-making; this can only be achieved if there is extensive collaboration between farmers, official and field veterinarians, poultry industry, the diagnostic laboratories, the epidemiology units and the central and local governments. Only in this way it will be possible to establish a network of collaboration able to make the best of the data and tools available in the effort to control avian influenza infections in poultry.

Acknowledgments

The authors wish to thank the staff of the Epidemiology and Virology Departments of the Istituto Zooprofilattico Sperimentale delle Venezie.

References

Capua, I. and Alexander, D.J., 2004. Avian influenza: recent developments. *Avian Pathology,* 33 (4), 393-404.

Capua, I., Marangon, S., Selli, L., et al., 1999. Outbreaks of highly pathogenic avian influenza (H5N2) in Italy during October 1997 to January 1998. *Avian Pathology,* 28 (5), 455-460.

Capua, I., Terregino, C., Cattoli, G., et al., 2003. Development of a DIVA (Differentiating Infected from Vaccinated Animals) strategy using a vaccine containing a heterologous neuraminidase for the control of avian influenza. *Avian Pathology,* 32 (1), 47-55.

CEC, 1992. Council Directive 92/40/EEC of 19 May 1992 introducing Community measures for the control of avian influenza. *Official Journal of the European Commission* (L 167, 22/06/1992), 1-16.

Franciosi, C., D' Aprile, P.N., Alexander, D.J., et al., 1981. Influenza A virus infections in commercial turkeys in north east Italy. *Avian Pathology,* 10 (3), 303-311.

Meulemans, G., 1986. Status of avian influenza in Western Europe. *In:* Slemons, R.D. ed. *Proceedings of the second international symposium on avian influenza.* US Animal Health Association, Georgia Center for Continuing Education, The University of Georgia, USA, Athens, GA, 77-83.

Papparella, V., Fioretti, A. and Menna, L.F., 1994. The epidemiological situation of avian influenza in Italy from 1990 to 1993 in feral bird populations and in birds in quarantine. *In: Proceedings of the joint first annual meetings of the national newcastle disease and avian influenza laboratories of countries of the European Communities, Brussels 1993.* 19–21.

Papparella, V., Fioretti, A. and Menna, L.F., 1995. The epidemiological situation of avian influenza in Italy in 1994. *In: Proceedings of the joint second annual meetings of the national newcastle disease and avian influenza laboratories of countries of the European Union, Brussels 1994.* 14-15.

Petek, M., 1982. Current situation in Italy. *In: Proceedings of the first international symposium on avian influenza 1981.* Carter Composition Corporation, Richmond, VA, 31–34.

6

An overview of the 2002 outbreak of low-pathogenic H7N2 avian influenza in Virginia, West Virginia and North Carolina

D.A. Senne#, T.J. Holt##, B.L. Akey###

Abstract

During the spring and summer of 2002, an outbreak of low-pathogenic H7N2 avian influenza virus (AIV) infected 210 flocks of chickens and turkeys in Virginia, West Virginia and North Carolina, and caused the destruction of more than 4.7 million birds. Although no epidemiologic link was established, the virus was related to the H7N2 virus circulating in the live-bird market system (LBMs) since 1994. An avian-influenza Task Force (TF), comprised of industry, state and federal personnel, was utilized in the control programme. The use of good safety and biosecurity practices was emphasized by TF commanders. Carcass-disposal options, which included burial in sanitary landfills, incineration and composting, proved to be problematic and caused delays in depopulation of infected premises. Surveillance activities focused on once-a-week testing of dead birds from all premises, biweekly testing of all breeder flocks and pre-movement testing. Additional surveillance carried out in backyard flocks and local waterfowl did not detect the H7 virus or specific antibodies to the virus. The outbreak emphasized the need to establish effective biosecurity barriers between the LBMs and commercial poultry.

Avian influenza (AI) is a viral disease that can affect many species of wild and domestic birds, including poultry. The AI virus (AIV) is comprised of 15 subtypes based on differences in antigenic nature of the surface haemagglutinin (HA) protein and is classified, based on pathogenicity, into low-pathogenic (LPAI) and highly pathogenic (HPAI) viruses (Swayne and Halvorson 2003). The natural reservoirs of avian influenza virus (AIV) are migratory waterfowl and shorebirds (Kawaoka et al. 1988; Slemons et al. 1974). However, the live-bird market system (LBMs) has been recognized as a significant man-made reservoir of poultry-adapted AIV and has been implicated in several outbreaks of AIV in commercial poultry in the United States (Committee on Transmissible Diseases of Poultry and Other Avian Species 2002; Davison et al. 2003).

The highly pathogenic form of AI is extremely contagious and lethal, causing sudden death in poultry, often without any warning signs of infection. Mortality in

---

\# National Veterinary Services Laboratories, Veterinary Services, Animal and Plant Health Inspection Service, US Department of Agriculture, P.O. Box 844, Ames, Iowa 50010, USA. E-mail: dennis.senne@aphis.usda.gov

\#\# Eastern Regional Office, Veterinary Services, Animal and Plant Health Inspection Service, US Department of Agriculture, 920 Main Campus Drive 27606-5202, Suite 200, Raleigh, North Carolina, USA

\#\#\# New York State Department of Agriculture and Markets, 1 Winners Circle, Albany, New York,USA

R. S. Schrijver and G. Koch (eds.), Avian Influenza, 41–47.
© 2005 *Springer. Printed in the Netherlands.*

flocks infected with HPAI can often reach 100%. It has been documented that HPAI can evolve through the mutation of LPAI H5 or H7 precursor viruses after circulating for extended periods in unnatural hosts such as domestic poultry (Capua and Marangon 2000; Horimoto et al. 1995; Kawaoka, Naeve and Webster 1984; Webster 1998). Low-pathogenic strains of AI can also be highly contagious often resulting in subclinical infections, allowing the virus to spread undetected for a period of time.

In March 2002, a LPAI H7N2 virus similar to a strain of H7N2 virus known to be present in the LBMs in Northeast United States was found to be present in commercial poultry in Virginia, West Virginia and North Carolina. To reduce the possibility of the H7 virus mutating to HPAI, a control programme was implemented to eradicate the H7N2 virus from commercial poultry in the region. Agriculture authorities in Virginia initially took steps to control the H7N2 LPAI through diagnostic testing, quarantines, surveillance, and depopulation and disposal of infected poultry. However, the rapid increase in the number of positive cases quickly overwhelmed the State's capacity to manage the outbreak. Consequently, the Commonwealth of Virginia asked the USDA for assistance in controlling the outbreak. This paper will provide an overview of the outbreak and methods used to control the outbreak.

Keywords: avian influenza; outbreak; surveillance; bird disposal

The poultry industry at risk in the Shenandoah Valley and North Carolina

The Shenandoah Valley, located in Northwest Virginia, is situated between the picturesque Blue Ridge mountain range to the east and Shenandoah Mountains to the west. The Valley is approximately 20-30 miles wide and stretches nearly 100 miles, north to south. At the time of the outbreak of low-pathogenic H7N2 AI, there were over 1,000 premises and more than 56 million commercial turkeys and chickens present in the Valley. Of the 1,000 premises, there were approximately 400 premises each with broilers and meat turkeys, 175 broiler breeder flocks, 50 turkey breeder flocks and 3 table-egg layer flocks.

North Carolina shares its northern boarder with Virginia. The state produces about 700 million broilers, 40 million turkeys and 1.4 million turkey breeders annually. This production represents about 30% of the nation's turkey hatching eggs and ranks second in meat-turkey production. The high density of poultry in the Shenandoah Valley and North Carolina provided the ultimate challenge to regulatory officials to control a highly contagious disease such as AI.

Chronology of the outbreak in Virginia and West Virginia

Clinical signs of respiratory disease and a drop in egg production were first observed on March 7, 2002 in a turkey breeder flock near Harrisonburg, Virginia. A diagnosis of H7N2 LPAI was confirmed by the National Veterinary Services Laboratories (NVSL), Ames, Iowa on March 12. One day prior to the appearance of clinical disease, some birds were moved from the affected house to another location for forced molting. Within the next few days, clinical disease was observed in several additional turkey breeder premises owned by the same company. It is believed that movement of infected birds and the use of a common rendering truck to pick up dead birds were responsible for tracking the infection from one breeder's premises to another. On March 21 it became apparent that this was not just a localized outbreak when a turkey grow-out farm located 30 miles north of the index farm and belonging to a different company, was diagnosed as positive. By March 28, 20 positive flocks were identified.

By April 12, 2002, more than 60 flocks were positive, with about half of the positive flocks awaiting depopulation. The poultry companies in the Valley insisted on depopulation of positive flocks; therefore, the State of Virginia began issuing 24-hour destruction orders for positive flocks. The State of Virginia also requested USDA assistance and on April 14 a joint Task Force (TF) comprised of State, Federal and Industry representatives was established, with headquarters located in Harrisonburg. Because an official 'state of emergency' was never declared, either at the state or federal level, all TF activities were carried out under state authorities to quarantine and order depopulation of infected flocks without indemnity. This was the first time the federal government participated in the control of LPAI in the United States. The control of LPAI in the United States is currently the responsibility of the state governments.

One of the most successful activities initiated by the TF was the 'barrel surveillance' programme that accomplished 100% coverage of all commercial poultry flocks once each week. This programme was started during the last week of April, 2002 and continued throughout the outbreak. Producers were required, at prearranged times, to place up to 10 dead birds per house in sealed garbage cans at the end of the driveway for sampling by the TF. Tracheal-swab pools (from up to 5 birds) were collected for laboratory testing. This activity facilitated collection of samples without compromising on-farm biosecurity procedures, thus limiting the spread of disease through surveillance activities. This practice proved to be very effective in detecting positive flocks that were infected but not showing clinical signs or where there may have been under-reporting of clinical disease by producers.

The last positive case in Virginia was identified on July 2, 2002, four months after the first case was diagnosed, and the final quarantine of positive premises was lifted on October 9, 2002. A total of 197 flocks, representing approximately 20% of the 1000 area commercial poultry farms, were infected with the H7N2 virus. Approximately 4.7 million birds, or 8.4% of the estimated 56 million birds at risk, were destroyed to control the outbreak. Turkeys accounted for 78% of the positive farms and included 28 turkey breeder flocks and 125 commercial meat-turkey flocks. Twenty-nine chicken broiler breeder flocks, 13 chicken broiler flocks and 2 of the 3 chicken egg-layer flocks in the Valley were also infected.

In addition to the infected flocks in Virginia, one flock in West Virginia was infected with the H7N2 virus. The poultry industries in Virginia and West Virginia are contiguous and it is suspected the disease was introduced into West Virginia from Virginia.

Although the USDA approved the use of an autogenous killed H7N2 vaccine, it was not used in this outbreak primarily because of company and allied industry concerns related to negative impacts on trade and to facilitate rapid eradication of the H7N2 virus from commercial poultry, thus reducing the opportunity for virus mutation that could lead to increased virulence.

The source of infection for the index flock was never established. However, the H7N2 strain responsible for this outbreak was shown to be genetically identical to the strain that caused recent outbreaks in Pennsylvania and that has been found in the LBMs in the Northeast United States since 1994. To assess the likelihood that wild waterfowl or backyard birds could have introduced the infection into commercial poultry, surveillance was carried out on more than 90 backyard flocks and 300 resident Canada geese from 23 sites in close proximity to infected poultry farms. Surveillance of waterfowl and backyard flocks yielded no positive isolations or serologic evidence of H7N2; antibodies to H6N2 AI virus were detected in some geese.

Federal compensation payments totalling $52.65 million were paid to growers and owners for the birds that were destroyed and for cost of bird disposal. The payments were made based on 75% of the appraised market value of the birds. An additional $13.5 million was spent on operational expenses for the outbreak TF. However, figures upward of $149 million have been used to reflect the total negative impact of the outbreak on the poultry industry and allied industries.

Chronology of events in North Carolina

During March and April, 2002, a total of 12 premises with over 60,000 birds were diagnosed positive for AI H7N2 in North Carolina. Of the 12 positive flocks, three were commercial turkey and quail flocks and 9 backyard flocks. The index case was detected on March 6, when the North Carolina Department of Agriculture and Consumer Services was notified that a North Carolina turkey flock processed in Virginia was positive for AI antibodies in serum collected at slaughter. During the following two weeks, surveillance detected H7N2 infection in one additional turkey flock and one quail flock at a nearby shooting preserve. Seven premises were identified as having received birds from the positive quail farm and four were confirmed as being infected. Investigations showed that the owner of one of the positive trace-back farms made regular trips to markets in Pennsylvania to sell goats; this activity could have been a source for the H7N2 virus for the outbreak. All infected premises in North Carolina were depopulated by the state without federal assistance.

Task-Force operations

A Task Force (TF) was established in Harrisonburg, Virginia and served as a headquarters for approximately 200 personnel at any given time. The mission of the TF was to assist the state of Virginia in control efforts by identifying and eliminating foci of infection and preventing spread of disease. Priorities identified by TF commanders included safety of TF personnel as well as adherence to strict biosecurity measures. All TF personnel were required to receive training in proper safety and biosecurity procedures before being assigned to an activity unit. During the outbreak, approximately 800 people from various federal and state agencies rotated through the TF. Personnel came from 46 states and several foreign countries.

Laboratory tests and surveillance

At the beginning of the outbreak only two testing modalities were available, the agar-gel immunodiffusion antibody test (AGID) at the state laboratory in Harrisonburg, Virginia and virus isolation in embryonated chicken eggs at the USDA's National Veterinary Services Laboratories (NVSL) in Ames, Iowa. Because of the time delays inherent in both of these test methods, days for seroconversion to occur after infection of a flock and days for results from virus isolation, additional test methods were sought that would provide rapid results to aid in the management of the outbreak. This led to the rapid adoption of the Directigen^{TM} FLU A (Becton, Dickinson and Company, Sparks, Maryland) membrane-based antigen-capture immunoassay and the real-time reverse-transcriptase polymerase chain reaction (RRT-PCR) test. A flock was diagnosed positive if clinical signs consistent with AI infection (respiratory signs, drop in egg production etc.) were present along with at least one positive laboratory test result. In the absence of clinical signs, positive results on two different types of tests were required to designate a flock as positive.

Early in the outbreak, swab specimens were tested by three methods: Directigen^{TM} test, virus isolation and RRT-PCR. As the outbreak progressed, the increased number of samples generated created a severe strain on the state and federal laboratories leading to changes in testing procedures. Results of a comparison of the three virus/antigen detection methods on over 3,500 specimens showed that the sensitivity and specificity of the AI RRT-PCR was 95% and 99% respectively when compared to virus isolation at the submission (farm) level. The Directigen^{TM} test was shown to be 80% sensitive and 99% specific compared to virus isolation. Therefore, the decision was made to stop testing by virus isolation and test only by Directigen^{TM} at the Harrisonburg laboratory and RRT-PCR at the NVSL. This marked the first poultry-disease outbreak in the United States where a molecular method was used as the primary diagnostic test for an eradication programme. Midway through the outbreak, equipment for the RRT-PCR was purchased for the Harrisonburg laboratory and personnel trained so that rapid, sensitive monitoring and surveillance for AI could occur locally. Timeliness of test turnaround was found to be absolutely critical to the successful management of the outbreak.

Prior to the outbreak, Virginia and North Carolina routinely conducted AGID tests to detect antibodies to AI virus from chicken and turkeys at processing plants and from breeders as part of the monitoring programme for the National Poultry Improvement Plan (NPIP). No evidence of AIV was detected in surveillance samples preceding the outbreak.

Antibody surveillance for AI was significantly increased during the outbreak and following the outbreak. In addition to antibody surveillance, breeder flocks were tested for AI virus antigen by the Directigen^{TM} and RRT-PCR tests at regular intervals. All meat birds going to slaughter were required to be Directigen^{TM} and RRT-PCR-negative before being moved. Over 96,000 serum samples and 40,000 swab specimens were tested during the outbreak.

Methods of bird disposal

Bird disposal proved to be a major issue during the outbreak in Virginia. Public protests following the burial of the index flock in plastic-lined pits on the farm of origin prompted the State Department of Environmental Quality to stop this practice unless land owners recorded such burial pits on the property deed and agreed to install long-term monitoring wells. These requirements made burial on the farm an unacceptable option as a disposal method. As a result, alternative methods for disposal were used, including air-curtain incineration, burial in large sanitary landfills, and composting. The use of 'mega-landfills', those landfills with the capacity and equipment to handle thousands of tons of carcasses per day, proved to be the most economical, despite sometimes having to transport the birds over long distances. Following euthanasia with carbon dioxide gas, carcasses were placed in sealed, leak-proof trucks for transport to the landfills. Task Force personnel monitored cleaning and disinfection of vehicles carrying dead birds from infected premises and prior to leaving disposal sites. No evidence of disease-spread was attributed to transportation of carcasses to landfills or incinerators.

Outcomes and lessons learned

A number of lessons were learned from this outbreak and response. The H7N2 viruses isolated from commercial poultry in Virginia, North Carolina and West Virginia were shown to be genetically similar to the H7N2 virus that has been circulating in the live-bird market system in the Northeastern United States since 1994. Therefore, greater effort must be made to establish barriers between commercial poultry and the live-bird markets and their suppliers to prevent tracking AI virus from these sources into commercial production facilities.

Epidemiologic studies showed that the spread of the H7N2 virus was primarily by people, fomites and contaminated equipment. There was very little evidence of airborne spread. The transport of dead birds (daily mortality) from the farm to rendering facilities for disposal was an especially high-risk activity.

Every outbreak is unique, and as such, flexibility and creative decision-making will be needed to solve problems that may arise. In addition, environmental considerations will figure prominently in disposal options.

This outbreak showed that multiple state and federal agencies can work effectively with industry and producers to quickly stamp out an outbreak of a highly contagious disease. The biosecurity measures used and methods for sample collection, euthanasia and disposal in this control programme did not contribute to further spread of the virus to other geographic areas.

Finally, trade considerations do play an important role in determining response policies such as stamping out, vaccination and disposal.

References

Capua, I. and Marangon, S., 2000. The avian influenza epidemic in Italy, 1999-2000: a review. *Avian Pathology,* 29 (4), 289-294.

Committee on Transmissible Diseases of Poultry and Other Avian Species, 2002. Avian influenza. *In: 106th Annual meeting of the United States Animal Health Association, St. Louis, MO, 17-24 October 2002.*

Davison, S., Eckroade, R.J., Ziegler, A.F., et al., 2003. A review of the 1996-98 nonpathogenic H7N2 avian influenza outbreak in Pennsylvania. *Avian Diseases,* 47 (special issue), 823-827.

Horimoto, T., Rivera, E., Pearson, J., et al., 1995. Origin and molecular changes associated with emergence of a highly pathogenic H5N2 influenza virus in Mexico. *Virology,* 213 (1), 223-230.

Kawaoka, Y., Chambers, T.M., Sladen, W.L., et al., 1988. Is the gene pool of influenza viruses in shorebirds and gulls different from that in wild ducks? *Virology,* 163 (1), 247-250.

Kawaoka, Y., Naeve, C.W. and Webster, R.G., 1984. Is virulence of H5N2 influenza viruses in chickens associated with loss of carbohydrate from the hemagglutinin? *Virology,* 139 (2), 303-316.

Slemons, R.D., Johnson, D.C., Osborn, J.S., et al., 1974. Type-A influenza viruses isolated from wild free-flying ducks in California. *Avian Diseases,* 18 (1), 119-124.

Swayne, D.E. and Halvorson, D.A., 2003. Influenza. *In:* Swayne, D.E. ed. *Diseases of poultry.* 11th edn. Iowa State University Press, Ames, IA, 135-160.

Webster, R.G., 1998. Influenza: an emerging disease. *Emerging Infectious Diseases,* 4 (3), 436-441.

7

Effectiveness of control measures on the transmission of avian influenza virus (H7N7) between flocks

J.A. Stegeman#, A. Bouma#, A.R.W. Elbers##, M. van Boven###, M.C.M. de Jong### and G. Koch##

Abstract

On 28 February 2003 an epidemic of fowl plague started in The Netherlands, caused by a highly pathogenic avian influenza virus (HPAIV) of type H7N7. The epidemic started in the 'Gelderse Vallei', spread to adjacent areas and to the province of Limburg. During the epidemic, 255 flocks were diagnosed as infected. The epidemic was combated by stamping out infected flocks and pre-emptive culling of flocks within a 1-km radius. Moreover, screening and tracing activities were implemented to enhance the detection of infected flocks. In addition, a transportation ban was enforced. In a further stage of the epidemic, poultry-free buffer zones were created, contacts between different parts of the country were reduced by compartmentalization and large areas were depopulated of all poultry. In all, 1,000 commercial flocks were pre-emptively culled, in addition to over 17,000 flocks of smallholders.

In this study we quantified the transmission of HPAIV between flocks during different phases of the epidemic. To stop an epidemic, infected flocks should be detected and depopulated before they have infected on average more than one other flock. This average number of secondary infections caused by one infectious flock is called the reproduction ratio (R_h). Upon the implementation of the control measures in the Gelderse Vallei, R_h dropped from 5.0 (95% confidence interval (CI): 2.9-8.6) to 0.91 (95%CI: 0.39-2.13). Moreover, in Limburg, R_h dropped from 2.5 to 0.86 (95%CI: 0.28-2.68) after the control measures came into force. Apparently, the measures significantly reduced the transmission of the virus. However, because the 95% confidence intervals of R_h after the implementation of control measures include one, it is uncertain whether the implemented measures are really sufficient to eliminate the virus in an area with a high poultry density. Consequently, additional control measures should be considered.

Key words: avian influenza; fowl plague; transmission; control measures

Utrecht University, Faculty of Veterinary Medicine, Department of Farm Animal Health Yalelaan 7, 3584 CL Utrecht, The Netherlands. E-mail: j.a.stegeman@vet.uu.nl
Central Institute for Disease Control (CIDC-Lelystad), Wageningen University and Research Centre, Lelystad, The Netherlands
Animal Sciences Group, Wageningen University and Research Centre, Division of Infectious Diseases, Lelystad, The Netherlands

R. S. Schrijver and G. Koch (eds.), Avian Influenza, 49–55.
© 2005 *Springer. Printed in the Netherlands.*

Introduction

On 28 February 2003 a suspicion of fowl plague or highly pathogenic avian influenza (HPAI) was notified to the Dutch Veterinary authorities. In a layer flock, comprising 7,150 animals, the hens had refused food and water since 22 February. Shortly afterwards, mortality increased and by 28 February approximately 90% of the hens had died. This presumed index flock appeared to be the onset of a huge epidemic, caused by a highly pathogenic avian influenza virus (HPAIV) of type H7N7.

Already on the same day four other herds were reported as fowl-plague-suspect. The epidemic was combated by movement restrictions, stamping out infected flocks and pre-emptive culling of flocks in the neighbourhood of infected flocks. Nevertheless, 1,255 commercial flocks and 17,421 flocks of smallholders had to be depopulated. The total number of animals killed mounted up to 25.6 million. In addition, the virus was also transmitted to a considerable number of humans that had been in close contact with infected poultry. Sadly, one of these persons died.

Obviously, the Dutch veterinary authorities want to reduce the probability that an unhoped-for new introduction of HPAIV will develop in an epidemic of similar size in the future. Consequently, it is important to have quantitative knowledge of the effectiveness of the control measures that were implemented during this epidemic. To stop an epidemic, infected flocks should be detected and depopulated before they infect on average more than one other flock. The average number of secondary infections caused by one infectious flock is called the reproduction ratio (R_h). The value of R_h depends on the infectivity of infected flocks, the susceptibility of non-infected flocks and the contact structure between flocks. Control measures intervene at one or more of these items.

In this study we quantified the transmission of HPAIV between flocks during different phases of the 2003 epidemic of fowl plague in The Netherlands.

Materials and methods

Outbreaks

During the epidemic 255 outbreaks of fowl plague were diagnosed. In 241 of these outbreaks clinical disease was present and H7N7-HPAIV was detected by virus isolation, PCR or both. On the basis of data collected from the outbreaks, the time of virus introduction was estimated for each of the infected flocks. The data comprised the contacts of the flocks with previous outbreaks and the history of the disease in the flock. Most farmers registered the mortality within their flocks over time. In transmission experiments, Van der Goot et al. (2003) observed that contact-infected chicken died 4-8 days after the exposure to HPAIV. By use of this period the time of virus introduction could be estimated from the moment the mortality in a flock rose above the base line. The remaining 14 flocks were detected by serological screening. No clinical disease was present, but one or more sera were positive in the H7 Haemaglutination Inhibition Assay. Estimation of the time of virus introduction is not possible in this situation.

Control measures

Several measures were implemented in order to stop the virus circulation. To start with, infected flocks were depopulated. In addition, flocks within a radius of 1 km of an outbreak were culled pre-emptively. In that situation the goal was to establish culling within 48 hours after notification of the outbreak. Furthermore, upon detection of an outbreak, official veterinarians performed forward and backward tracing. Moreover, a transport ban was implemented; a protection zone and a surveillance zone were established around an infected flock; and the flocks in these areas were examined for clinical disease symptoms. As the epidemic proceeded, more measures were implemented such as the establishment of buffer zones, compartmentalization of The Netherlands into separate units and the culling of all poultry in large areas.

Analysis of data

The data were analysed by a model that has previously been used to quantify the transmission of Classical Swine Fever virus between pig farms (Stegeman et al. 1999) and the transmission of foot-and-mouth disease virus between cattle farms (Bouma et al. 2003). In short, this model assumes that the number of newly infected flocks (C) in a time interval dt can be described as

$$C = \beta * I \quad (1)$$

with β, the infection rate parameter (the average number of new infections per infectious flock per time unit) and I, the number of infectious flocks present in time interval dt. Consequently, if we have for each time unit the number of newly infected flocks and the number of infectious flocks, β can be established by a Generalized Linear Model (GLM) with a Poisson distribution and a log-link function. In this model C is the dependent variable and the natural logarithm of I is included as an offset variable. Moreover, we included different periods of the epidemic as explanatory variables. These periods were: 1) period until detection of an outbreak in a previously unaffected area; 2) first two weeks after detection of the first case in an area (assuming that it would take some time before measures were fully implemented); 3) remaining time until the end of the epidemic in the region. Finally, the established βs were transformed into R_hs, using the equation

$$R_h = \beta * T \quad (2)$$

with T = the duration of the infectious period. We assumed that a flock was not infectious for other flocks during the first two days upon virus introduction.

Results

Descriptive epidemiology

The epidemic started in the middle of The Netherlands, in a region called the 'Gelderse Vallei'. This region has a very high density of poultry flocks. Figure 1 shows the course of the number of outbreaks throughout the epidemic. Figure 2 shows the distribution of the infected flocks over the Netherlands.

The number of diagnosed outbreaks increased during the first week of March 2003. Subsequently, the number of outbreaks per day fluctuated between 2 and 11 until the end of March, without a clear trend up or down. During that period, outbreaks were only detected in the Gelderse Vallei. By the beginning of April, however, the number of outbreaks per day dropped markedly. The reason was that almost all flocks in the Gelderse Vallei had been culled at that time and consequently, the number of susceptible flocks became restricted. However, after that drop, the number of outbreaks fluctuated throughout most of April between 0 and 5 outbreaks per day. The reason was that by the end of March the virus had escaped from the Gelderse Vallei to the Southern part of The Netherlands. Here the virus also continued to spread in the province of Limburg (Figure 2). Again, until 15 April no clear trend up or down is visible in the epidemic curve and the total number of outbreaks in Limburg mounted

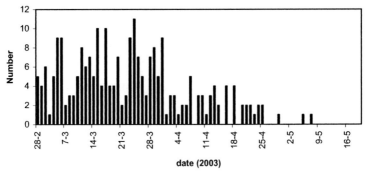

Figure 1. Course of the number of outbreaks (based on virus isolation or positive PCR) detected per day during the 2003 epidemic of fowl plague in The Netherlands

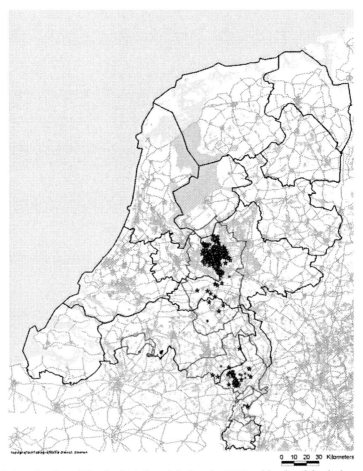

Figure 2. Localization of the flocks infected during the 2003 epidemic of fowl plague in The Netherlands. Infected flocks are represented as stars; suspected flocks, not confirmed at that time, are represented as dots

up to 43. Next, however, the epidemic faded out, with only a few outbreaks in May. By then, in the Southern part of The Netherlands large areas had been emptied of all commercial poultry.

The distribution of the infected and culled flocks over the various types of poultry is shown in Table 1. If we assume that the total number of flocks culled per poultry type represents the flocks at risk for infection, laying flocks had a much higher probability of getting fowl plague than broiler flocks (Relative Risk (RR) = 10.5 [95% Confidence Interval (95%CI): 2.7-41.5]), duck flocks (RR = 5.7 [95%CI: 1.5-22.4]), and hobby flocks (RR = 334 [99%CI: 191-585]). However, the probability of turkey flocks and breeder flocks to become infected was comparable to the probability of laying flocks.

Type of poultry	Number of infected flocks (%)	Number of culled flocks
Layers	168 (25%)	673
Turkeys	18 (29%)	62
Breeders	34 (31%)	110
Broilers	2 (2.4%)	84
Ducks	2 (4.3%)	46
Pets	13 (0.07%)	17,421
Others	4 (1.4%)	280

Table 1. Distribution of the number of infected flocks (based on virus isolation or positive PCR) and culled flocks over the various types of poultry

Between-flock transmission

The transmission was analysed separately for the Gelderse Vallei and adjacent areas and Limburg in the Southern part of The Netherlands. In both regions, there was no significant difference in β for the first two weeks after the first outbreak had been detected in the region and the period after those first two weeks. Therefore, in each of the two regions only two periods were distinguished, 1) before detection of the first outbreak and 2) after detection of the first outbreak. The results of the analysis are shown in Table 2.

	Before implementation of control measures		After implementation of control measures	
	β^a	$R_h{}^b$	β	R_h
Gelderse	0.42	5.0	0.17	0.91
Vallei	(0.27-2.41)	(2.9-8.6)	(0.11-0.28)	(0.39-2.13)
Limburg	0.42	2.9	0.18	0.86
	(0.17-3.65)	(no CI)	(0.08-0.46)	(0.28-2.68)

[a] β, infection-rate parameter, average number of new infections per infectious flock per day
[b] R_h, reproduction ratio, average number of new infections per infected flock95% confidence intervals between brackets.

Table 2. Infection-rate parameter (β) and reproduction ratio (R_h) in the Gelderse Vallei and Limburg, before and after the implementation of measures to eliminate HPAIV

The control measures caused a significant (p=0.0002) reduction of the transmission of HPAIV between flocks in the Gelderse Vallei. In Limburg a similar reduction of β was observed, although in this region the difference between the period before detection of the first outbreak and after that detection was not significant (p=0.07).

53

However, the total number of outbreaks in this area, and thus the power of the comparison, was limited in comparison to the Gelderse Vallei, Moreover, the between-flock transmission in both areas shows a remarkable similarity, in the period before detection of the first outbreak as well as in the period afterwards.

In the Gelderse Vallei, the average infectious period decreased from 11.8 days (95% confidence interval (CI) 10.1-13.4) of the 5 herds that were notified as fowl-plague-suspect on the first day of the epidemic to 6.0 (95% CI 5.5-6.5) days in the first week and 5.2 (95% CI 4.9-5.5) during the rest of the epidemic. In Limburg, the estimated infectious period of the flock first affected was 7 days. In the subsequent period of the epidemic the infectious period was 4.8 (95% CI 4.3-5.3) days.

Although the between-flock transmission decreased upon implementation of the control measures, the accompanying R_h was only slightly less than 1 in both regions. The latter implies that, even after implementation of the control measures, we cannot falsify the hypothesis $R_h > 1$. Moreover, the reduction in the infectious period after the first week in the Gelderse Vallei did not result in a significant reduction in the value of R_h.

Discussion

In this study we quantified the transmission of HPAIV-H7N7 between poultry flocks during different phases of the 2003 epidemic of fowl plague in The Netherlands. In the Gelderse Vallei, the region first affected, control measures markedly reduced the between-flock transmission of HPAIV. Unfortunately, we are unable to establish the contribution of each of the individual measures to the reduction of the virus transmission. Nevertheless, it is clear that the stamping-out policy markedly contributed to this reduction, because it halved the infectious period and thus also halved R_h. Moreover, pre-emptive culling may also have caused some of the reduction of β, if infected flocks have been culled in a very early stage of the infection when it was still impossible to detect HPAIV. However, measures like the movement ban, restricted admittance of persons to farms and other hygienic measures probably also contributed to the reduction of the transmission.

Remarkably, in Limburg the values of β before and after the implementation of control measures were almost the same as in the Gelderse Vallei. Consequently, the awareness of HPAIV being in the country did not result in a lower infection rate. However, this awareness probably caused a shorter infectious period of the flock first infected in Limburg as compared to the flocks first infected in the Gelderse Vallei, and thus in a lower value of R_h. As a result, the number of infected flocks was much smaller at the start of the epidemic in Limburg than in the Gelderse Vallei.

In the Gelderse Vallei as well as in Limburg the transmission rate was not significantly reduced further during later phases of the epidemic. This raises questions as to the effectiveness of control measures implemented in these later phases, such as compartmentalization, creation of poultry-free buffer zones and the culling of all poultry in large areas. These questions are raised even more, because it is likely that the between-flock transmission towards the end of the epidemic was underestimated, because of the strong reduction in the number of susceptible flocks in the affected areas.

Upon the implementation of the control measures, in both regions the value of R_h was only slightly below one. Consequently, a considerable number of flocks still became infected during the epidemic. Moreover, the 95% CIs of R_h include one, indicating that it is uncertain whether the applied set of control measures is really capable of eliminating the virus. Also here we should bear in mind that the transmission was probably underestimated due to the reduction of the number of susceptible flocks. Consequently, new or additional measures should be considered in future epidemics. Further reduction of the infectious period seems difficult. Consequently, we should focus on measures to reduce β, such as further restriction of the contacts between flocks or vaccination.

In conclusion, we found that a package of control measures including stamping out infected herds, pre-emptive culling of surrounding flocks, movement bans, screening and tracing and hygienic measures markedly reduced the transmission of HPAIV H7N7. However, it is uncertain whether this reduction is really sufficient to eliminate the virus in an area with a high poultry density.

References

Bouma, A., Elbers, A.R.W., Dekker, A., et al., 2003. The foot-and-mouth disease epidemic in The Netherlands in 2001. *Preventive Veterinary Medicine,* 57 (3), 155-166.

Stegeman, A., Elbers, A.R.W., Smak, J., et al., 1999. Quantification of the transmission of classical swine fever virus between herds during the 1997-1998 epidemic in The Netherlands. *Preventive Veterinary Medicine,* 42 (3/4), 219-234.

Van der Goot, J.A., De Jong, M.C.M., Koch, G., et al., 2003. Comparison of the transmission characteristics of low and high pathogenicity avian influenza A virus (H5N2). *Epidemiology and Infection,* 131 (2), 1003-1013.

VACCINATION

8

Currently available tools and strategies for emergency vaccination in case of avian influenza

I. Capua[#] and S. Marangon[##]

Abstract

Recent epidemics of highly contagious animal diseases included in the list A of the OIE such as foot-and-mouth disease, classical swine fever and avian influenza (AI) have led to the implementation of stamping-out policies resulting in the depopulation of millions of animals. The enforcement of a control strategy based on culling of animals that are infected, suspected of being infected or suspected of being contaminated, which is based only on the application of sanitary restrictions on farms, may not be sufficient to avoid the spread of infection, particularly in areas that have high animal densities, thus resulting in mass depopulation.

In the European Union, the directive that imposes the enforcement of a stamping-out policy (92/40/EC) for AI was adopted in 1992 but was drafted in the 1980s. The poultry industry has undergone substantial changes in the last twenty years, mainly resulting in shorter production cycles and greater animal densities per territorial unit. Due to these organizational changes, infectious diseases are significantly more difficult to control as a result of the greater number of susceptible animals reared per given unit of time and the difficulties in applying adequate biosecurity measures.

The slaughter and destruction of great numbers of animals is also questionable from an ethical point of view, particularly when human-health implications are negligible. For this reason, mass depopulation has raised serious concerns for the general public and has recently led to very high costs and economic losses for the national and federal governments, the stakeholders and ultimately for the consumers.

In the past, the use of vaccines in such emergencies has been limited by the impossibility of differentiating vaccinated/infected from vaccinated/non-infected animals. The major concern was that through trade or movement of apparently uninfected animals or products, the disease could spread further or might be exported to other countries. For this reason export bans have been imposed on countries enforcing a vaccination policy.

This paper takes into account the possible strategies for the control of avian influenza infections, bearing in mind the new proposed definition of AI. In detail, an overview of the advantages and disadvantages of using conventional inactivated (homologous and heterologous) vaccines and recombinant vaccines is presented and

---

[#] OIE and National Reference Laboratory for Newcastle Disease and Avian Influenza, Istituto Zooprofilattico Sperimentale delle Venezie Viale dell'Università 10, 35020, Legnaro, Padova, Italy. E-mail: icapua@izsvenezie.it
[##] Centro Regionale di Epidemiologia Veterinaria, Istituto Zooprofilattico Sperimentale delle Venezie, Viale dell'Università 10, 35020, Legnaro, Padova, Italy
59

R. S. Schrijver and G. Koch (eds.), Avian Influenza, 59–74.
© 2005 *Springer. Printed in the Netherlands.*

discussed. Reference is made to the different control strategies including the restriction measures to be applied in case of the enforcement of a vaccination policy. In addition, the implications of a vaccination policy on trade are discussed.

In conclusion, if vaccination is accepted as an option for the control of AI, vaccine banks including companion diagnostic tests must be established and made available for immediate use.

Keywords: avian influenza; vaccination; intervention strategies; poultry

Introduction

Recent epidemics of highly contagious animal diseases included in the list A of the Office International des Epizooties (OIE) such as foot-and-mouth disease, classical swine fever and avian influenza (AI) have led to the implementation of stamping-out policies resulting in a depopulation involving millions of animals. The implementation of a control strategy based on culling of animals that are infected, suspected of being infected or suspected of being contaminated, which is based only on the application of sanitary restrictions, may not be sufficient to avoid the spread of infection. This event is particularly foreseeable in areas that have high animal densities, and inevitably results in mass-depopulation policies. There is an increased risk of disease spread in these areas and the financial consequences of any occurring epidemic are severe (Capua and Marangon 2000; Dijkhuizen and Davies 1995; Gibbens et al. 2001; Meuwissen et al. 1999).

With reference to AI, the EU directive that imposes the enforcement of a stamping-out policy (92/40/EC) was adopted in 1992 but was drafted in the '80's (CEC 1992). The poultry industry has undergone substantial changes in the last twenty years, mainly resulting in shorter production cycles and in the development of densely populated poultry areas (DPPA). As a result of these organizational changes, infectious diseases are significantly more difficult to control due to the greater number of susceptible animals reared per given unit of time and to the difficulties in applying adequate biosecurity programmes. In order to avoid the destruction of great numbers of animals, the possibility of pursuing different control strategies should be considered.

Until recent times highly pathogenic avian influenza (HPAI) was considered a rare disease in domestic poultry with only 17 episodes being reported worldwide in the 40-year period 1959-1998. However, three further outbreaks have occurred since 1999, resulting in 11 outbreaks since 1991 and six in the six years covering 1997-2003. Recently, there also appears to have been a marked increase in the number of low-pathogenicity AI (LPAI) outbreaks caused by H5 and H7 viruses. The countries, subtypes and approximate number of birds involved are listed in Table 1. From 1997 to date 14 significant outbreaks due to H5 or H7 subtypes have been reported in poultry, three of which have had human-health implications. The approximate number of birds culled for AI in the past 6 years has been 63 million, with hundreds of millions of birds involved (Capua and Alexander 2004).

The slaughter and destruction of great numbers of animals is also questionable from an ethical point of view. For this reason, mass depopulation has raised serious concerns from the general public. The policy has also led to very high costs and economical losses for the Community budget, the Member States, the stakeholders and ultimately for the consumers.

Table 1. Outbreaks of LPAI and HPAI caused by H5 and H7 viruses in recent years

Country	Year[s]	Subtype	Virulence	Approximate no. of birds infected/culled	Control
Mexico Guatemala, El Salvador	1994-2003 2000 2001	H5N2	LPAI/HPAI	>1.000,000,000	Vaccination
Pennsylvania	1996-1998	H7N2	LPAI	2,623,116	Depopulation
Australia	1997	H7N4	HPAI	310,565	Stamping out
Hong Kong	1997-2003	H5N1	HPAI	~3,000,000	Stamping out Vaccination
Italy	1997	H5N2	HPAI	7741	Stamping out
Ireland	1998	H7N7	LPAI	320,000	Depopulation
N. Ireland	1998	H7N7	LPAI	?	Depopulation
Italy	1998	H5N9	LPAI	2,000	Stamping out
Belgium	1999	H5N2	LPAI	100	Stamping out
Italy	1999-2001	H7N1	LPAI HPAI LPAI	17,000,000	Stamping out Vaccination + stamping out
Germany	2001	H7N7	LPAI	145	Stamping out
Pakistan	2001	H7N3	HPAI/LPAI	>10,000,000?	Vaccination
USA (NC/VA)	2002	H7N2	LPAI	~5,000,000	Stamping out
Chile	2002	H7N3	LPAI/HPAI	~1,000,000	Stamping out
Italy	2002-2003	H7N3	LPAI	>6,000,000	Vaccination + stamping out
The Netherlands Belgium Germany	2003	H7N7	HPAI	30,283,000 2,700,000 419,000	Stamping out
USA (CT)	2003	H7N2	LPAI	2,900,000	Vaccination

In the EU, the possibility of vaccinating to aid control policies in such emergencies has been limited by the inability to differentiate vaccinated–infected from vaccinated–non-infected animals. The major concern was that through trade or movement of vaccinated animals or their products, the disease could spread further or might be exported to other countries, primarily because it was not possible to establish whether the vaccinated animals had been field-exposed.

The following paper takes into account the recent developments in vaccinology, which may represent valid tools for the control of avian-influenza infections, bearing in mind the new definition of AI proposed by the EU (Document Sanco/B3/AH/R17/2000) and by the OIE (*ad hoc* expert group on Avian Influenza, Animal Health Code Commission meeting of 29-30 October 2002) and the possibility of enforcing an emergency vaccination programme with the products currently available. Reference will be made to the type of vaccines available, the efficacy of these vaccines, their limitations and the possibility of identifying infected animals in a vaccinated population.

Definition of avian influenza

Avian influenza viruses all belong to the *Influenzavirus A* genus of the *Orthomyxoviridae* family and are negative-stranded, segmented RNA viruses. The influenza-A viruses can be divided into 15 subtypes on the basis of the haemagglutinin (H) antigens. In addition to the H antigen, influenza viruses possess one of nine neuraminidase (N) antigens. Virtually all H and N combinations have been isolated from birds, thus indicating the extreme antigenic variability that is a hallmark of these viruses. Changes in the H and N composition of a virus may be brought about by genetic reassortment in host cells. One of the consequences of genomic segmentation is that if co-infection by different viruses occurs in the same cell, progeny viruses may originate from the reassortment of parental genes originating from different viruses. Thus, since the influenza-A virus genome consists of 8 segments, from two parental viruses 256 different combinations of progeny viruses may arise theoretically.

Current EU legislation (CEC 1992) defines avian influenza as "an infection of poultry caused by any influenza-A virus which has an intravenous pathogenicity index in six-week-old chickens greater than 1.2 or any infection with influenza-A viruses of H5 or H7 subtype for which nucleotide sequencing has demonstrated the presence of multiple basic amino acids at the cleavage site of the haemagglutinin". However it has been proved that highly pathogenic avian influenza (HPAI) viruses emerge in domestic poultry from low-pathogenicity (LPAI) progenitors of the H5 and H7 subtypes. It therefore seems logical that HPAI viruses and their LPAI progenitors must be controlled, when they are introduced in domestic poultry populations (Scientific Committee on Animal Health and Animal Welfare 2000). The new proposed definition of AI for the OIE and the EU (Scientific Committee on Animal Health and Animal Welfare 2000) is "an infection of poultry caused by either any influenza A virus which has an intravenous pathogenicity index in 6-week-old chickens greater than 1.2 or any influenza A virus of H5 or H7 subtype". With reference to the present paper, the term avian influenza applies to all avian influenza viruses of the H5 and H7 subtype, regardless of their virulence and of their pathogenicity for domestic poultry.

Rationale behind the use of vaccines

When an outbreak of avian influenza occurs in an area with a high population density, the application of rigorous biosecurity measures might not be possible. In this case the disease may spread very rapidly, and the tracing and culling capacities might not be adequate. This results in enormous efforts being spent on chasing the disease rather than on actively imposing a barrier to its spread. The only additional measure that can be taken to attempt to reduce disease spread is vaccination. The expected results of the implementation of a vaccination policy on the dynamics of infection are primarily those of reducing the susceptibility to infection (i.e. a higher dose of virus is necessary for establishing productive infection) and reducing the amount of virus shed into the environment. The association between a higher infective dose necessary to establish infection and less virus contaminating the environment represents a valuable support to the eradication of infection.

The efficacy of an emergency vaccination programme is inversely correlated with the time span between the diagnosis in the index case and the implementation of mass vaccination. For this reason, it is imperative that if emergency vaccination is considered as a possible option in a given country, vaccine banks must be available in the framework of national contingency plans.

It should be clear that vaccination can be used for a variety of different scopes. It can be used to protect birds and reduce spread in case an HPAI epidemic is out of control and there is the risk of spread to other DPPAs. When there is no more evidence of virus circulation the vaccinated birds can be culled or dealt with appropriately. It can also be used to support eradication measures for LPAI in a DPPA. This is a longer-term strategy, which may also imply trade of commodities, provided they come from vaccinated animals that have not been field-exposed.

Currently available vaccines

Conventional vaccines
Inactivated homologous vaccines

These vaccines were originally prepared as 'autogenous' vaccines, i.e., vaccines that contain the same avian-influenza strain as the one causing the problems in the field. They have been used extensively in Mexico and Pakistan during the AI epidemics (Swayne and Suarez 2000).

The efficacy of these vaccines in preventing clinical disease and in reducing the amount of virus shed in the environment has been proven through field evidence and experimental trials (Swayne and Suarez 2000). The disadvantage of this system is the impossibility of differentiating vaccinated from field-exposed birds unless unvaccinated sentinels are kept in the shed. However, the management (identification, bleeding and swabbing) of sentinel birds during a vaccination campaign is time-consuming and rather complicated since they are difficult to identify, and they may be substituted with seronegative birds in the attempt to escape restrictions imposed by public health officials.

Inactivated heterologous vaccines

These vaccines are manufactured in a similar way to the previous ones. They differ in the fact that the virus strain used in the vaccine is of the same H type as the field virus but has a heterologous neuraminidase. Following field exposure, clinical protection and reduction of viral shedding are ensured by the immune reaction induced by the homologous H group while antibodies against the neuraminidase induced by the field virus can be used as a marker of natural infection (Capua and Marangon 2000).

For both homologous and heterologous vaccines, the degree of clinical protection and the reduction of shedding are improved by a higher antigen mass in the vaccine (Swayne et al. 1999). For heterologous vaccines the degree of protection is not strictly correlated to the degree of homology between the haemagglutinin genes of the vaccine and challenge strains (Swayne and Suarez 2000). This is definitely a great advantage because it enables the establishment of vaccine banks since the master seed does not contain the virus that is present in the field and may contain an isolate (preferably of the same lineage) available before the epidemic.

Recombinant vaccines
Several recombinant fowlpox virus expressing the H5 antigen have been developed (Beard, Schnitzlein and Tripathy 1992; Beard, Schnitzlein and Tripathy 1991; Swayne et al. 2000; Swayne, Beck and Mickle 1997; Webster et al. 1996) and one has

been licensed and is currently being used in Mexico (Swayne and Suarez 2000). Experimental data have also been obtained for fowlpox-virus recombinants expressing the H7 antigen (Boyle, Selleck and Heine 2000). Other vectors have been used to deliver successfully the H5 or H7 antigens such as constructs using infectious laryngotracheitis virus (ILTV) (Lüschow et al. 2001).

The only field experience with a recombinant virus to control AI has been obtained in Mexico (Villareal-Chavez and Rivera Cruz 2003), where it has been used in the vaccination campaign against an LPAI H5N2 virus.

No such product has been licensed in the EU to date.

Trade implications

Until recent times, vaccination against avian influenza viruses of the H5 and H7 subtypes, was not considered or practised in developed countries since it implied export bans on live poultry and on poultry products (CEC 1994). In case of an infection with an H5 or H7 virus, regardless of the virulence of the isolate, export bans have also been imposed. Export bans frequently represent the major cause of economic loss due to OIE List A diseases.

Whilst the severe clinical signs caused by HPAI ensure a prompt diagnosis and facilitate the implementation of a stamping-out policy, the inconspicuous nature of the disease caused by viruses of low pathogenicity make this infection difficult to diagnose. Detection of infection is only possible with the implementation of appropriate surveillance programmes. Bearing in mind the new proposed definition of AI, and the potential mutation of LPAI of the H5 and H7 subtypes to HPAI it is easy to understand why these bans have been imposed. For the sake of trade, freedom from AI should be demonstrated in a given country or compartment by ongoing surveillance programmes. This approach is supported by the fact that in several recent outbreaks, infection with a virus of low pathogenicity was only detected once infection was widespread, and often out of control.

In absence of vaccination, trade bans imposed on a given area last until freedom from infection can be demonstrated in the affected population. In the case of the adoption of a vaccination policy that does not enable the application of a 'DIVA' (differentiating infected from vaccinated animals) strategy (either for the type of vaccine used or because the monitoring system in place does not guarantee that infection is no longer circulating) this also results in prolonged trade bans. On the contrary, if it is possible to demonstrate that the infection is not circulating in the vaccinated population trade bans may be lifted.

Such 'marker' vaccination strategies offer attractive control options for OIE List-A diseases. In case of an outbreak of avian influenza in a DPPA the option of vaccinating should be pursued. To safeguard international trade a control strategy that enables the differentiation between vaccinated–infected and vaccinated–non-infected animals should be implemented. The possibility of using vaccines would support restriction-based control measures, thus reducing the risk of a major epidemic and the subsequent mass stamping-out policy.

Options for control

It is extremely difficult to establish fixed rules for the control of infectious diseases in animal populations, due to the unpredictable number of variables involved.

However, with reference to AI, some basic scenarios may be hypothesized, and on the basis of the considerations made above some guidelines may be drawn, which are reported in Table 2.

Table 2. Guidelines for the application of control policies for AI

H5/H7 virus pathogenicity	Index case flock	Evidence of spread to industrial circuit	Population density in area	Policy
HPAI/LPAI	Backyard	No	High/Low	Stamping out
HPAI/LPAI	Backyard	Yes	Low	Stamping out
			High	Vaccination
HPAI/LPAI	Industrial	No	High/Low	Stamping out
HPAI/LPAI	Industrial	Yes	Low	Stamping out
			High	Vaccination

There are several crucial steps that must be planned for if avian influenza represents a risk. Firstly the index case must be promptly identified. This should not represent a problem if the virus is of high pathogenicity, but it can be a serious concern if the virus if of low pathogenicity. For this reason countries or areas at risk of infection should implement specific surveillance systems to detect infection with LPAI as soon as it appears.

Secondly, a timely assessment must be performed of whether there has been spread to the industrial poultry population of that area. This is a crucial evaluation, which must be made available for decision-makers.

Once an AI outbreak has been identified eradication measures based on the stamping out or controlled marketing of slaughter birds on infected farms must be enforced. The choice between these two options must be taken bearing in mind the pathogenicity and transmissibility of the virus, the density of poultry farms around the affected premises, the economical value of the affected birds, the logistics for slaughter/stamping out and the collaborative approach of farmers/producers. With reference to the Italian experience a stamping-out policy was generally applied to LPAI-infected young meat birds, breeders and layers, while controlled marketing was applied for older meat birds approaching slaughter age. This strategy enables the reduction of the restriction periods (i.e. if infected young turkeys, breeders or layers were kept on the farms the restriction period could last several months) and hence facilitates faster restocking.

In addition, restriction measures on the movement of live poultry, vehicles and staff must be imposed in the areas at risk.

Finally, if vaccination is the proposed strategy, vaccine banks should be available for immediate use and a contingency plan must be enforced. A territorial strategy must be implemented. It must include restriction measures (Tables 3 and 4) and an ongoing set of adequate controls (Figure 1) that enable public authorities to establish whether the virus is circulating or not in the vaccinated population and assess the efficacy of the vaccination programme.

Table 3. Basic restriction and monitoring measures to be enforced on the movements of live poultry and poultry products originating from and/or destined for farms or plants located in the vaccination area (VA)

Commodity	Restrictions to movements towards the VA	Restrictions to movements inside the VA	Restrictions to movements outside the VA
Hatching eggs	- shall be transported directly to the hatchery of destination - (and their packaging) must be disinfected before dispatch - tracing-back of egg lots in the hatchery shall be guaranteed	- must originate from a vaccinated or unvaccinated breeding flock that has been tested, with negative results, according to Table 4 - shall be transported directly to the hatchery of destination - (and their packaging) must be disinfected before dispatch - tracing-back of egg lots in the hatchery shall be guaranteed	- must originate from a vaccinated or unvaccinated breeding flock that has been tested, with negative results, according to Table 4 - shall be transported directly to the hatchery of destination - (and their packaging) must be disinfected before dispatch - tracing-back of egg lots in the hatchery shall be guaranteed
Day-old chicks	must be destined for a poultry-house where: - no poultry is kept - cleansing and disinfection operations have been carried out	- must originate from hatching eggs satisfying the conditions mentioned above - must be destined for a poultry house where no poultry is kept and where cleansing and disinfection operations have been carried out	- must originate from hatching eggs satisfying the conditions mentioned above - must be destined for a poultry house where no poultry is kept and where cleansing and disinfection operations have been carried out
Ready-to-lay pullets	must be: - housed in a poultry house where no poultry has been kept for at least 3 weeks, and cleansing/disinfection operations have been carried out - vaccinated at the farm of destination	must: - have been vaccinated regularly against avian influenza - have been tested, with negative results, according to Table 4 - be destined for a farm located in the VA and housed in a poultry house where no poultry has been kept for at least 3 weeks, and cleansing/disinfection operations have been carried out - be officially inspected within 24 hours before loading - be virologically and serologically tested with negative results before loading (sentinel birds)	must: - not have been vaccinated - have been tested, with negative results, according to Table 4 - be destined for a poultry house where no poultry has been kept for at least 3 weeks, and cleansing/disinfection operations have been carried out - be officially inspected within 24 hours before loading - be virologically and serologically tested with negative results before loading

Commodity	Restrictions to movements towards the VA	Restrictions to movements inside the VA	Restrictions to movements outwards the VA
Poultry for slaughter	- must be sent directly to the abattoir for immediate slaughter - must be transported by lorries that operate, on the same day, only on farms located outside the VA - lorries must be washed and disinfected under official control before and after each transport	- shall undergo a clinical inspection within 48 hours before loading - must be directly sent to the abattoir for immediate slaughter - must be serologically tested before loading - the abattoir must guarantee that accurate washing and disinfection operations are carried out under official supervision - shall be transported by lorries that operate, on the same day, only on farms located inside the VA - lorries must be washed and disinfected before and after each transport	- shall undergo a clinical inspection within 48 hours before loading - must be sent directly to an abattoir designated by the competent veterinary authority for immediate slaughter - must be serologically tested before loading - the abattoir must guarantee that accurate washing and disinfection operations are carried out under official supervision - shall be transported by lorries that operate, on the same day, only on farms located inside the VA - lorries must be washed and disinfected before and after each transport
Table eggs	must be: - sent directly to a packaging centre or a thermal-treatment plant designated by the competent authority - transported using disposable packaging materials that can be effectively washed and disinfected	must: - originate from a flock that has been tested, with negative results, as laid down in Table 4 - be sent directly to a packaging centre or a thermal-treatment plant designated by the competent authority - be transported using disposable packaging material or packaging material that can be effectively washed and disinfected	must: - originate from a flock that has been tested, with negative results, as laid down in Table 4 - be sent directly to a packaging centre or a thermal-treatment plant designated by the competent authorities - be transported using disposable packaging material or packaging material that can be effectively washed and disinfected

67

Table 4. Basic restrictions to be applied to the trade of fresh meat produced from poultry originating from the vaccination area (VA)

Commodity	Unrestricted to international trade	Restricted to national trade
Fresh poultry meat	- originating from birds vaccinated against avian influenza with a heterologous subtype vaccine can be dispatched to other countries, provided that the meat comes from slaughter-turkey flocks that: (i) have been regularly inspected and tested with negative results for avian influenza as follows. For the testing of: — vaccinated animals, the anti-N discriminatory test shall be used — sentinel animals, either the haemagglutination-inhibition test (HI), the AGID test or the ELISA test shall be used. However, anti-N discriminatory test shall also be used if necessary (ii) have been clinically inspected by an official veterinarian within 48 hours before loading. Sentinel animals shall be inspected with particular attention (iii) have been serologically tested with negative results with the iIFA test (iv) must be sent directly to a slaughterhouse designated by the competent authority and be slaughtered immediately on arrival - produced from poultry not vaccinated against avian influenza and originating from the VA	originating from holdings located in the VA cannot be dispatched to other countries, if produced from poultry: (i) vaccinated against avian influenza with a homologous subtype vaccine (ii) vaccinated against avian influenza with a heterologous subtype vaccine and not tested, with negative results, using the anti-N discriminatory test (iii) originating from seropositive poultry flocks subjected to controlled marketing (iv) coming from poultry holdings located in the restriction zone (minimum 3 km radius) that must be established around any LPAI-infected farms for at least two weeks

MONITORING MEASURES IN THE VACCINATION AREA

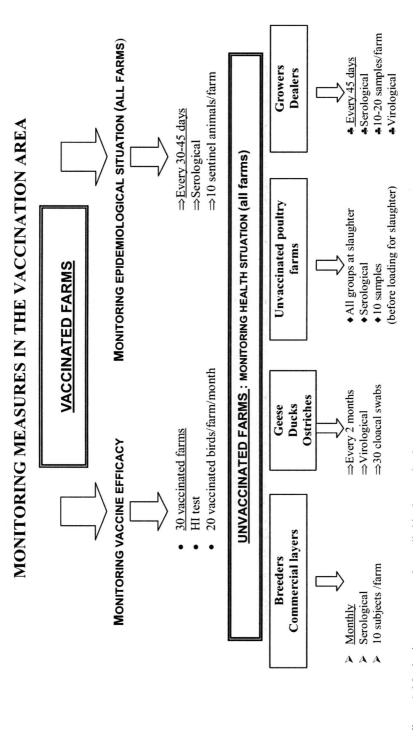

VACCINATED FARMS

MONITORING VACCINE EFFICACY

- 30 vaccinated farms
- HI test
- 20 vaccinated birds/farm/month

MONITORING EPIDEMIOLOGICAL SITUATION (ALL FARMS)

⇒ Every 30-45 days
⇒ Serological
⇒ 10 sentinel animals/farm

UNVACCINATED FARMS : MONITORING HEALTH SITUATION (all farms)

| Breeders Commercial layers | Geese Ducks Ostriches | Unvaccinated poultry farms | Growers Dealers |

Breeders Commercial layers

⋏ Monthly
⋏ Serological
⋏ 10 subjects /farm

Geese Ducks Ostriches

⇒ Every 2 months
⇒ Virological
⇒ 30 cloacal swabs

Unvaccinated poultry farms

◆ All groups at slaughter
◆ Serological
◆ 10 samples
(before loading for slaughter)

Growers Dealers

♣ Every 45 days
♣ Serological
♣ 10-20 samples/farm
♣ Virological

Figure 1. Monitoring measures to be applied in the vaccination area

Applications in the field

Conventional vaccines
Inactivated homologous vaccines

These products have recently been used in the attempt to control avian-influenza infections in Pakistan and Mexico (Swayne and Suarez 2000), but under those specific conditions they have not been successful in eradicating the infection. Conversely, in one instance, in Utah (Frame et al. 1996), the use of this vaccination strategy has been successful. The reason for the discrepancy of the results probably lies in the efficacy of the direct control measures, which must be implemented to support a vaccination campaign.

Inactivated heterologous vaccines

This vaccination strategy has been used successfully over the years in Minnesota (Halvorson 2002); however, in these instances vaccination was never implemented to control infections caused by viruses of the H5 or H7 subtypes. In addition the heterologous neuraminidase was not used as a marker of infection.

Conversely, in Italy during 2000-2002 this strategy was used to supplement control measures for the eradication of the H7N1 LPAI virus (CEC 2000). In order to control the re-emergence of LPAI virus and to develop a novel control strategy, a co-ordinated set of measures, including strict biosecurity, a serologic monitoring programme and a 'DIVA' strategy were enforced (Commission Decision 2001/721/CE as amended). The 'DIVA' strategy was based on the use of an oil-emulsion-based inactivated vaccine with the same haemagglutinin and a heterologous neuraminidase (N) subtype from the field virus, in this case an H7N3 strain.

The possibility of using the diverse N groups to differentiate between vaccinated and naturally infected birds, was achieved through the development of an *ad hoc* serological test to detect the specific anti-N1 antibodies (Capua et al. 2003).

Control of the field situation was achieved through an intensive sero-surveillance programme aimed at the detection of the LPAI virus through the regular testing of sentinel birds in vaccinated flocks and through the application of the anti-N1-antibody detection test. Serological monitoring was also enforced in unvaccinated flocks, located both inside and outside the vaccination area. In addition, the efficacy of the vaccination schemes was evaluated in the field through regular serological testing of selected flocks.

After the first year of vaccination, the epidemiological data collected indicated that the H7N1 virus was not circulating. This was considered sufficient by the EU Commission to lift the marketing restrictions on fresh meat obtained from vaccinated poultry provided that animals had been tested with negative results using the discriminatory test (Commission Decision 2001/847/CE) (CEC 2001).

It is clear that due to the unpredictable nature of the epidemiology of this disease, which could result in the introduction of other avian-influenza subtypes, this solution is to be considered 'tailored' for a given epidemic.

Recombinant vaccines

The only field experience with these vaccines has been obtained in Mexico, where it has been used in the vaccination campaign against the H5N2 virus. Avian influenza has not been eradicated in Mexico, probably because an eradication programme based on a territorial strategy and including monitoring and restriction was not established.

Recombinant live vectored vaccines also enable the differentiation between infected and vaccinated birds, since they do not induce the production of antibodies against the nucleoprotein antigen, which is common to all AI viruses. Therefore, only field-infected birds will exhibit antibodies to the AGP or ELISA test directed towards the detection of group A (nucleoprotein) antibodies.

Since these vaccines have encountered some difficulties in licensing, their use is restricted to countries in which they are legally available. In addition, these vaccines will not replicate, and induce protective immunity, in birds that have had field exposure to the vector (i.e. fowlpox or infectious laryngotracheitis viruses) (Lüschow et al. 2001; Swayne, Beck and Kinney 2000). Since serological positivity to these viruses is widespread (due to field exposure and vaccination) in the poultry population, and can be in some instances unpredictable, the use of these vaccines is limited to a population which is seronegative to the vector virus. In addition, the use of these vaccines is restricted to species in which the vector virus will replicate. For example, ILTV will not replicate in turkeys, and since these birds are particularly important in the epidemiology of AI, the use of this vaccine is limited to areas in which turkeys are not present.

Discussion

From the data presented it appears that emergency vaccination is a sensible option if there is evidence of the introduction of a highly transmissible AI virus in a densely populated poultry area, or whenever the epidemiological situation indicates that there could be massive and rapid spread of infection. In addition, emergency vaccination should be considered where applicable, when birds of high economic value (e.g. pedigree flocks) or rare (endangered) birds are at risk of infection. It is clear that vaccination represents a tool to support eradication, and will be a successful tool only if coupled with restriction and increased biosecurity.

Considering the advantages and disadvantages of the products and diagnostic tools taht are currently available, if no recombinant products are licensed in that country, heterologous vaccination rather than homologous vaccination should be practiced in case of an emergency. The main reason for this would be that it would enable the differentiation between vaccinated and naturally exposed birds, through the development/application of an appropriate test. At present only the anti-neuraminidase-based test is available and has been validated. In our opinion however, this test represents a starting point on which future developments of the 'DIVA' strategy can be based. The development of novel candidate vaccines and of additional tests that enable the detection of field infection in vaccinated populations should be a priority for pharmaceutical industries and for research institutions, since for all the reasons listed above vaccination is already an option for the control of avian influenza.

If the country has access to licensed recombinant products, the use of these vaccines is acceptable taking into consideration the immune status of the population against the vector, since seropositivity impedes the replication of the vector virus and therefore the establishment of immunity. The issue of the replicating capacity of the vector in different species must also be addressed.

In conclusion, recent events including devastating epidemics in densely populated poultry areas, public-health concerns on animal-welfare issues and the introduction of novel technology into vaccinology have encouraged consideration of alternative control strategies for OIE List-A diseases which were unthinkable of only a few years ago. This has also been supported by the development of reliable, sensitive and specific diagnostic companion tests. Countries, areas and enterprises at risk of infection should imperatively enforce surveillance programmes and have contingency plans in case of a disease outbreak, which may include vaccination. If the latter is considered an option, among other issues, the contingency plan must foresee the establishment of licensed vaccine banks that enable the 'DIVA' strategy, thus safeguarding animal health, animal welfare and international trade.

Acknowledgements

The support of Dr. Manuela Dalla Pozza, CREV, and of Dr. Maria Elizabeth Pittman, EU Commission is gratefully acknowledged.

References

Beard, C.W., Schnitzlein, W.M. and Tripathy, D.N., 1991. Protection of chickens against highly pathogenic avian influenza virus (H5N2) by recombinant fowlpox viruses. *Avian Diseases,* 35 (2), 356-359.

Beard, C.W., Schnitzlein, W.M. and Tripathy, D.N., 1992. Effect of route of administration on the efficacy of a recombinant fowlpox virus against H5N2 avian influenza. *Avian Diseases,* 36 (4), 1052-1055.

Boyle, D.B., Selleck, P. and Heine, H.G., 2000. Vaccinating chickens against avian influenza with fowlpox recombinants expressing the H7 haemagglutinin. *Australian Veterinary Journal,* 78 (1), 44-48.

Capua, I. and Alexander, D.J., 2004. Avian influenza: recent developments. *Avian Pathology,* 33 (4), 393-404.

Capua, I. and Marangon, S., 2000. The avian influenza epidemic in Italy, 1999-2000: a review. *Avian Pathology,* 29 (4), 289-294.

Capua, I., Terregino, C., Cattoli, G., et al., 2003. Development of a DIVA (Differentiating Infected from Vaccinated Animals) strategy using a vaccine containing a heterologous neuraminidase for the control of avian influenza. *Avian Pathology,* 32 (1), 47-55.

CEC, 1992. Council Directive 92/40/EEC of 19 May 1992 introducing Community measures for the control of avian influenza. *Official Journal of the European Commission* (L 167, 22/06/1992), 1-16.

CEC, 1994. Commission Decision 1994/438/EC of 7 June 1994 laying down the criteria for classifying third countries and parts thereof with regard to avian influenza and Newcastle Disease in relation to imports of fresh poultry meat amending decision 93/342/EEC. *Official Journal of the European Commission* (L181), 35-43.

CEC, 2000. Commission Decision 2000/721/EC of 7 November 2000 on introducing vaccination to supplement the measures to control avian influenza in Italy and on specific movement control measures. *Official Journal of the European Commission* (L291), 33-36.

CEC, 2001. Commission Decision 2001/847/CE of 30 November 2001 amending for the third time Decision 2000/721/EC to modify the Italian avian influenza vaccination programme and current trade restrictions for fresh meat originating from vaccinated turkeys. *Official Journal of the European Commission* (L315), 61-63.

Dijkhuizen, A.A. and Davies, G., 1995. *Animal health and related problems in densely populated livestock areas of the Community: proceedings of a workshop held in Brussels, 22 - 23 November 1994.* Office for Official Publications of the E.C., Luxembourg. Report / EUR 16609 en.

Frame, D.D., McCluskey, B.J., Buckner, R.E., et al., 1996. Results of an H7N3 avian influenza vaccination program in commercial meat turkeys. *In: Proceedings 45th western poultry disease conference, Cancun, Mexico, May 1-5, 1996.* 32.

Gibbens, J.C., Sharpe, C.E., Wilesmith, J.W., et al., 2001. Descriptive epidemiology of the 2001 foot-and-mouth disease epidemic in Great Britain: the first five months. *Veterinary Record,* 149 (24), 729-743.

Halvorson, D.A., 2002. The control of H5 or H7 mildly pathogenic avian influenza: a role for inactivated vaccine. *Avian Pathology,* 31 (1), 5-12.

Lüschow, D., Werner, O., Mettenleiter, T.C., et al., 2001. Protection of chickens from lethal avian influenza A virus infection by live-virus vaccination with infectious laryngotracheitis virus recombinants expressing the hemagglutinin (H5) gene. *Vaccine,* 19 (30), 4249-4259.

Meuwissen, M.P., Horst, S.H., Huirne, R.B., et al., 1999. A model to estimate the financial consequences of classical swine fever outbreaks: principles and outcomes. *Preventive Veterinary Medicine,* 42 (3/4), 249-270.

Scientific Committee on Animal Health and Animal Welfare, 2000. *The definition of avian influenza and The use of vaccination against avian influenza.* European Commission, Scientific Committee on Animal Health and Animal Welfare. [http://europa.eu.int/comm/food/fs/sc/scah/out45_en.pdf]

Swayne, D.E., Beck, J.R., Garcia, M., et al., 1999. Influence of virus strain and antigen mass on efficacy of H5 avian influenza inactivated vaccines. *Avian Pathology,* 28 (3), 245-255.

Swayne, D.E., Beck, J.R. and Kinney, N., 2000. Failure of a recombinant fowl poxvirus vaccine containing an avian influenza hemagglutinin gene to provide consistent protection against influenza in chickens preimmunized with a fowl pox vaccine. *Avian Diseases,* 44 (1), 132-137.

Swayne, D.E., Beck, J.R. and Mickle, T.R., 1997. Efficacy of recombinant fowl poxvirus vaccine in protecting chickens against a highly pathogenic Mexican-origin H5N2 avian influenza virus. *Avian Diseases,* 41 (4), 910-922.

Swayne, D.E., Garcia, M., Beck, J.R., et al., 2000. Protection against diverse highly pathogenic H5 avian influenza viruses in chickens immunized with a recombinant fowlpox vaccine containing an H5 avian influenza hemagglutinin gene insert. *Vaccine,* 18 (11/12), 1088-1095.

Swayne, D.E. and Suarez, D.L., 2000. Highly pathogenic avian influenza. *Revue Scientifique et Technique,* 19 (2), 463-482.

Villareal-Chavez, C. and Rivera Cruz, E., 2003. An update on avian influenza in Mexico. *In: Proceedings of the 5th International Symposium on Avian Influenza, Georgia Center for Continuing Education, The University of Georgia, Athens, Georgia, USA.* American Association of Avian Pathologists, 783-1267. Avian Diseases vol. 47 (3 Suppl.).

Webster, R.G., Taylor, J., Pearson, J., et al., 1996. Immunity to Mexican H5N2 avian influenza viruses induced by a fowl pox-H5 recombinant. *Avian Diseases,* 40 (2), 461-465.

9

Evaluation of vaccination to support control of H5N1 avian influenza in Hong Kong

T.M. Ellis[#], L.D. Sims, H.K.H. Wong, L.A. Bissett, K.C. Dyrting, K.W. Chow, C.W. Wong

Abstract

In 1997, 2002 and 2003 highly pathogenic avian influenza (HPAI) was diagnosed on chicken farms in Hong Kong. Following the February-April 2002 outbreak, vaccination using a killed oil-adjuvanted H5N2 avian influenza vaccine was evaluated as an additional control measure on 22 farms within a 2-km radius of the four farms that were depopulated following infection with HPAI H5N1 virus. Vaccination produced satisfactory flock antibody responses. The serological response was improved following a second dose of vaccine and the response to vaccination was poorer when delivered to older birds compared to birds first vaccinated at 8 days of age. Infection with field virus was not detected in any of these vaccinated flocks so the protective effect of the vaccine was tested under secure laboratory conditions on vaccinated and unvaccinated chickens challenged with HPAI H5N1 virus. Vaccinated birds were protected from disease, virus excretion was not detected in eight of ten vaccinated birds and the two birds that did excrete virus excreted much less virus than unvaccinated controls (> 1000 fold reduction). In December 2002 HPAI H5N1 outbreaks in 2 waterfowl parks and deaths in wild water birds in Hong Kong were followed by outbreaks on five previously unvaccinated chicken farms. Vaccination used in the face of outbreaks on three of these farms, coupled with selective culling, resulted in elimination of H5N1 virus infection from these farms. These investigations showed that the killed H5N2 vaccine, used in conjunction with enhanced biosecurity measures on chicken farms and in poultry markets, reduced the risk of H5N1 avian-influenza outbreaks in Hong Kong and consequently the risk of spread to humans.
Keywords: avian influenza H5N1; killed H5N2 vaccine; chickens; Hong Kong; evaluation

Introduction

Outbreaks of H5N1 highly pathogenic avian influenza (HPAI) have occurred in Hong Kong in chickens and other gallinaceous poultry in 1997, 2001, 2002 and 2003. High mortality rates were seen in gallinaceous birds on farms (1997, 2002, 2003) and/or in poultry markets (1997, 2001, 2002, 2003) in all outbreaks, but not in wild or captive water birds until late 2002. Outbreaks of H5N1 HPAI occurred in waterfowl (geese, ducks and swans) and other wild water birds (Little Egrets *Egretta garzetta*, and in captive Greater Flamingo *Phoenicopterus ruber*) at two waterfowl parks in

---

[#] Corresponding author: Dr T.M. Ellis, Agriculture, Fisheries and Conservation Department, 303 Cheung Sha Wan Road, Kowloon, Hong Kong SAR. E-mail: ellis_trevor@afcd.gov.hk

R. S. Schrijver and G. Koch (eds.). Avian Influenza. 75–84.
© 2005 *Springer. Printed in the Netherlands.*

Hong Kong in December 2002. HPAI H5N1 virus was also isolated from two dead wild Grey Heron (*Ardea cinerea*) and a Black–headed Gull (*Larus ridibundus*).

In the 1997 avian-influenza outbreak a strain of H5N1 HPAI virus spread directly to humans, causing 18 influenza cases with death in six people (Shortridge et al. 2000). This had a dramatic effect on the perception of avian influenza worldwide and a substantial economic impact in Hong Kong. The commercial chicken population of Hong Kong, (1.3 million birds) was killed in December 1997 (Shortridge 1999), live poultry markets remained closed for several months, there was a significant drop in tourism to Hong Kong and a comprehensive H5N1 testing and surveillance system had to be introduced for local and imported poultry. The 2001 retail poultry market H5N1 HPAI outbreak resulted in culling of 440,000 birds in poultry markets, closure of the markets and culling of 800,000 unaffected older market-age chickens on farms. The H5N1 HPAI outbreaks on farms in early 2002 resulted in the culling of 900,000 chickens. Although no human cases of disease occurred, disruption to the poultry trade and effects on tourism to Hong Kong (due to perceived risks by tourists) caused significant economic and social costs.

After the 2001 outbreak the poultry farm and market biosecurity measures and monitoring systems in place since 1998 were further enhanced. Detailed epidemiological study of the February-April 2002 H5N1 HPAI outbreak by an Investigation Team recommended further measures to improve farm and market biosecurity. All these measures have now been included in individual farm biosecurity plans for all local poultry farms and these form part of the poultry farms' licence conditions. Due, in part, to the large daily movement of poultry into Hong Kong from Southern China and the possibility of H5N1 virus infections occurring in the region the H5 avian-influenza vaccine was introduced as an additional control measure.

Vaccines have been used in other countries to assist in the control of avian influenza. Countries using vaccines against influenza viruses include Italy (Capua et al. 2003a), USA (Halvorson 2002), Mexico (Villarreal and Flores 1997) and Pakistan (Naeem 1997). Mostly vaccination has been directed against low-pathogenic strains of avian influenza virus but Mexico and Pakistan have successfully used vaccine against highly pathogenic H5 or H7 avian influenza viruses. Experimental studies have shown that commercially available H5 avian-influenza vaccines could protect poultry from 1997 Hong Kong strains of H5N1 HPAI virus (Swayne et al. 2001).

Field evaluation of a commercial killed H5N2 vaccine on chicken farms in Hong Kong in terms of adequate H5 antibody response and protection from challenge with current Hong Kong H5N1 HPAI viruses is reported. The effect of vaccination in the face of an outbreak was also examined in three chicken farms which were not included in the original vaccination evaluation trials.

Materials and methods

Evaluation procedures for an H5N2 avian-influenza vaccine in Hong Kong

The vaccine used in these evaluations was Nobilis® Influenza H5, an inactivated avian-influenza Type A H5N2 virus (A/Chicken/Mexico/232/94/CPA) water-in-oil emulsion vaccine (Intervet International, Boxmeer, The Netherlands). The dose, 0.5 ml, was administered subcutaneously into the neck in young chickens and intramuscularly into the breast muscle of older chickens by farm workers after instruction from Agriculture, Fisheries and Conservation Department (AFCD) staff.

Initially (Phase-1 vaccination programme), an evaluation trial was conducted on chicken farms in the district involved in the last four cases of H5N1 HPAI in the

February-April 2002 outbreak. After H5N1 HPAI outbreaks in waterfowl parks and wild birds in December 2002 the vaccination programme was extended to farms considered at higher risk from wild-bird transmission (Phase 2). Subsequently, with H5N1 HPAI outbreaks on five unvaccinated chicken farms, vaccination was used during the outbreak on three farms and its effect was evaluated (Phase 3).

Chicken farms in Hong Kong are broiler farms rearing yellow meat chickens favoured by consumers in Hong Kong. They receive day-old chickens from breeder farms in Mainland China. Chickens are reared in cages and are marketed as a batch at around 90-100 days of age. There were about 150 active chicken farms in Hong Kong at the time, producing 8 million chickens per year (approximately 20% of consumption) for sale in Hong Kong. The remaining birds are derived from the Mainland.

Phase-1 vaccination programme

Phase-1 vaccine evaluation was conducted on 22 chicken farms in the Pak Sha district of the New Territories in Hong Kong and located within 2 km of the last four chicken farms infected with H5N1 virus in the February-April 2002 outbreak. Vaccinations commenced in April 2002 and the programme continued on these farms until March 2003. In the first round all chickens between 8 and 55 days of age were vaccinated and revaccinated 4 weeks later. Subsequently, all new batches of chickens were vaccinated at 8-10 days of age and again four weeks later. Each batch had a group of 30 individually identified chickens left unvaccinated (sentinels).

Blood was collected from the 30 unvaccinated sentinel chickens and 30 individually identified vaccinated chickens 4 weeks after the first and second dose of vaccine and again within 5 days of sale. Serum was tested by standard haemagglutination-inhibition test (Alexander 2000) for antibody to H5 avian influenza using avian-influenza A/chicken/Hong Kong/97 (H5N1) virus antigen. Sentinel or vaccinated chickens that died were subjected to necropsy examination and tested for the presence of avian influenza virus by standard procedures (Alexander 2000). Prior to sale a sample of 60 chickens per batch had cloacal swabs collected and tested for presence of H5 virus by NASBA (Collins et al. 2002) or real-time RT-PCR (Spackman et al. 2002).

Statistical analysis was conducted on post-vaccination antibody responses on the vaccinated flocks from these farms using analysis of covariance with repeated measures (Wong 2003).

The criteria empirically set for successful use of the vaccine in this farmer-administered vaccination programme were that at least 90% of batches developed a measurable antibody response (HI titre ≥ 16) after one vaccination, that $\geq 70\%$ of chickens per batch had a HI titre ≥ 16 after two doses of vaccine and that the geometric mean titre (GMT) by HI test of the batch after two doses was ≥ 20.

Experimental challenge procedure

An experimental challenge with a current H5N1 virus was conducted in a biosafety level 3 laboratory using 71-day-old vaccinated and sentinel chickens from a single batch of chickens from one vaccinated farm. The virus used was a H5N1 ("Z" genotype) virus isolated from a dead chicken from a retail poultry market in April 2002. The virus was inoculated via the eye, nose and beak into 10 vaccinated chickens and 10 sentinel chickens such that each chicken was challenged with approximately 15,000 ($10^{4.2}$) egg infectious doses (EID) of virus per chicken.

After challenge all chickens were examined daily for signs of disease and swabs were collected daily from cloaca and throat of all surviving chickens for the duration of the trial (10 days). Surviving chickens were tested for antibody at 10 days post-challenge.

Phase-2 vaccination programme

In late December 2002 a decision was taken to extend the H5N2 vaccination programme to areas around the initial 22 farms and to farms deemed to have a high potential risk of exposure to wild birds (Phase-2 vaccination programme). This was done in response to waterfowl wild-bird and retail poultry-market outbreaks.

Vaccination commenced on 53 farms from 23 December 2002. As with the Phase-1 (Pak Sha) vaccination trial, initially all chickens up to 55 days of age were vaccinated twice at a monthly interval and then subsequent batches of chickens were vaccinated at 8 and 36 days of age. Antibody response was measured a month after the second vaccination to determine the proportion of batches with $\geq 70\%$ of chickens with HI antibody titre ≥ 16 and the overall percentage of chickens with HI antibody titre ≥ 16.

Phase 3 – vaccination of chicken flocks during an outbreak

In late December 2002 to end of January 2003 outbreaks of HPAI caused by H5N1 virus occurred on five previously unvaccinated chicken farms. Immediate quarantine and movement control was initiated and two farms were completely depopulated. On three farms affected sheds were depopulated and strict biosecurity procedures, vaccination and close daily monitoring of other sheds and surrounding farms started. In two farms (TKP1 and TKP2) infection spread to adjacent sheds before vaccination had time to induce an immune response. This enabled monitoring of the effect of the vaccine in the face of field challenge with virulent H5N1 virus.

On farm TKP1, HPAI was detected in eight sheds (30,000 chickens) in relatively close proximity between 7 and 16 January 2003. These were depopulated in two stages, three on 11 January and five on 16 January. A single large, more isolated shed (20,000 birds) that had been vaccinated 9 days previously was closely monitored.

Farm TKP2 was adjacent to TKP1 and various sheds had been vaccinated between 8 and 14 January 2003 as part of the control programme around TKP1. Low-level mortality was detected in two adjacent sheds on TKP2 on 20 January 2003 and these sheds (5,300 birds) were depopulated immediately with close monitoring of other sheds.

Sequential measurements of serum H5 antibody levels and cross-sectional virus culture of cloacal and throat swabs from chickens in affected sheds were conducted on TKP1 until 14 February 2003 (38 days p/v) and on TKP2 until 20 February 2003 (42 days p/v) using procedures referred to above. Subsequently all ongoing batches from these farms were checked for absence of H5 antibody in unvaccinated sentinel chickens and virus-testing using RRT-PCR for H5 virus was conducted using procedures referred to above on 60 chickens per batch until May 2003.

A third farm (SKT) situated over 1 Km away had one affected shed (5,600 chickens) on 20 January 2003 that was depopulated next day. The other sheds on the farm and nine nearby farms were vaccinated on 23 January and all were subjected to daily monitoring for H5N1 HPAI-affected birds. Unvaccinated sentinel chickens were tested for H5 antibody and 60 random cloacal swabs were tested for virus from each batch of market-aged chickens.

Results

Antibody responses and virus monitoring on Phase-1 farms
In total 248 batches of chickens involving 1.35 million birds were vaccinated and fully tested, by 31 March 2003. No clinical outbreaks of disease associated with H5N1 virus were detected on any of these vaccinated farms. Nor was any H5N1 virus detected in tests conducted on the chickens from these farms prior to sale or on dead sentinel chickens. Ninety-eight percent of batches had detectable antibody after the first dose of vaccination and 80% of the 248 batches of chickens developed satisfactory antibody levels after two doses of vaccination. The monthly breakdown of results for field vaccinations is shown in Table 1.

Table 1. Results of Phase-1 vaccination in Pak Sha based on the month when chickens received the first dose of vaccine

Month given first dose	Number of successful batches	Total batches vaccinated	Proportion successful batches (%)	Mean no. antibody-positive *	Mean antibody titre **
April 2002	21	42	50%	70%	31.0
May 2002	14	21	67%	76%	37.7
June 2002	17	27	65.4%	70%	39.3
July 2002	17#	23	73.9%	82%	49.3
Aug. 2002	16	17	94.2%	86%	50.6
Sept. 2002	23	25	92%	87.7%	75.2
Oct. 2002	25	28	89.3%	84.8%	69.6
Nov. 2002	25	29	86.2%	84.4%	66.4
Dec. 2002	16	20	80.0%	77.5%	43.4
Jan. 2003	13	16	81.3%	85.2%	53.8
Total	188	248	75.8%	80.5% +	50.9 +

* % of birds with HI titre $\geq$ 16 at one month after second vaccination
** Mean geometric titre of batches at one month after second vaccination
\# Five of these batches were very marginally below the success target
+ Weighted mean allowing for numbers of batches per month

Of the 60 batches that did not meet the success targets, 16 batches (first vaccinated in April 2002) were vaccinated as older chickens (>21 days), 15 batches were marginally outside targets, 12 batches occurred on three farms on which batches of birds vaccinated in the early part of the trial responded poorly but the response in later batches improved subsequently, and 17 were individual batches within farms that otherwise had a good response to vaccination in other batches of chickens.

Statistical analysis of the antibody-response data confirmed that the GMT and the proportion of seropositive chickens after two vaccinations were significantly higher than after one vaccination (P< 0.001 for both). Older birds had a significantly decreased GMT (P = 0.007) and a significantly reduced percentage of seropositive birds (P< 0.001) after second vaccination. Vaccination of birds early in the programme was also associated with significantly lower GMT after second vaccination than vaccination later in the programme (P< 0.001). The latter is most probably related to proportions of flocks vaccinated at an older age in the first one to two months of the programme. By the third month all birds were being vaccinated at 8-10 days and 36-38 days of age. There was also a significant association (P = 0.001) between farm of origin and GMT after second vaccination (Wong 2003).

In total 202 dead sentinel or vaccinated chickens from 67 batches of birds from the vaccine trial farms were necropsied and cultured for H5N1 virus. Most deaths involved individual sentinel or vaccinated chickens in the 20-50-day-old range. Several farms had more than four sentinel deaths in a batch on more than one occasion caused by virulent Newcastle disease virus (ND). Known immunosuppressive diseases diagnosed in necropsied chickens included infectious bursal disease (IBD), ND, Marek's disease, infectious laryngo-tracheitis and chronic respiratory disease (CRD) complex (infectious bronchitis virus +/or ND virus + *Mycoplasma* spp.). Other parasitic and bacterial diseases encountered included coccidiosis, airsacculitis, peritonitis, coryza and colibacillosis.

Experimental challenge

The H5 HI titres in the vaccinated chickens at challenge ranged from 32 to 256. No sentinel chickens had antibody to H5 influenza virus. After challenge, none of the vaccinated chickens became ill whereas all the unvaccinated sentinel chickens died within 3 days. This indicated a highly significant level of protection (Chi-square = 20, p = 0.00000). Vaccinated chickens excreted markedly reduced titres of virus via cloaca or throat. Throat swabs from unvaccinated chickens contained an average of 18,000 EID of virus. No vaccinated chickens had detectable virus in throat or cloaca on day 2 or any other day except for day 4 post-infection when two vaccinated chickens had low levels of virus in the throat ($10^{1.0}$ EID) (see Table 2).

Table 2. Summary of results of experimental H5N1 challenge

Type of chicken	Proportion infected (mean days to 1st signs)	Proportion dead (mean death time)	HI antibody-positive 10 days post-challenge	Cloacal virus (titre on day 2)*	Throat virus (titre on day 2)*
Not vaccinated	10/10 (2)	10/10 (2.3)	None survived	7/10 ($10^{2.03}$)	10/10 ($10^{4.25}$)
Vaccinated (GMT = 119)	0/10	0/10	10/10 (GMT= 239)	0/10	0/10**

* Virus titre = Log_{10} embryo infectious doses (EID) per swab via chicken embryo allantoic route.
** virus was detected in two vaccinated birds at low level ($10^{1.0}$ EID) on day 4 only.

Phase-2 vaccination programme

A total of 60 batches of chickens received two doses of vaccine. Thirty-two batches (53.3%) had $\geq$ 70% of chickens with HI antibody titre $\geq$ 16 and overall 69.2% of the chickens had HI antibody titre $\geq$ 16.

No clinical outbreaks of disease associated with H5N1 virus infection were detected on any of the Phase-2 vaccinated farms, no H5N1 virus was detected in tests conducted on any dead or sick chickens submitted from these farms as part of the sentinel and surveillance programme and none of the sentinel chickens gave positive H5 antibody results. Some 1.55 million vaccine doses were given to chickens on these farms with no adverse effects apparent from use of this killed vaccine.

The vaccination response in Phase 2 was not significantly different from that for the first round of Phase 1, which also involved vaccination of birds ranging from 8 to 55 days of age. In Phase 2 nearly half (28/60) the batches received their first dose of vaccine when older than 21 days of age and only 19 batches received their first vaccination at 8-10 days of age. Problems of vaccination technique with older chickens and immunosuppressive diseases are likely to have had a similar effect on the first round of vaccination on these farms as they did with the Phase 1 farms.

Phase 3 – vaccination of chicken flocks during an outbreak

Use of the killed H5N2 vaccine in the face of a H5N1 outbreak on Farms TKP1 and TKP2 showed that it was able to provide significant protection from disease and shut down virus excretion by 13-18 days post-vaccination. On the third farm (SKT) where vaccination was used and in the nine vaccinated surrounding farms no H5N1 HPAI or H5 virus infection was detected by serological monitoring of sentinels or virus culture of 60 random cloacal swabs per batch. In the latter farms it is possible that by rapid depopulation of the affected shed and by strict attention to biosecurity other sheds on the farm or surrounding farms may not have been exposed to H5N1 virus.

Discussion and conclusions

The vaccine evaluation studies showed that the killed H5N2 vaccine produced a satisfactory flock antibody response against H5 haemagglutinin antigen and could protect vaccinated chickens against highly pathogenic avian influenza that is caused by current Hong Kong strains of H5N1 virus. Moreover, vaccination produced a substantial reduction (>1000-fold) in excretion of infectious H5N1 virus in vaccinated compared with unvaccinated chickens and was able to protect chickens and shut down the virus excretion on infected farms by 13-18 days post-vaccination.

Post-immunization antibody measurement in humans vaccinated with A/New Caledonia/20/99 (H1N1) vaccines showed 78% of adult and 66% of elderly vaccinees responded, and for recent human isolates, including H1N2 viruses, the responses were similar with 70% of adults and 55% of elderly vaccinees positive. Vaccines containing influenza A/Panama/2007/99 (H3N2) and recent similar H3N2 viruses gave antibody responses in 66-71% of adult and 72-78% of elderly vaccines (Studies with inactivated influenza virus vaccines 2003). These population antibody responses are in line with the empirical success criteria used in these trials ($\geq$ 70% of the flock antibody-positive with GMT of $\geq$ 20). The overall result of 80.5% antibody-positive birds with GMT of 50.9 supports our conclusion that the vaccine produced a satisfactory flock antibody response. While no studies have been conducted to determine the level of herd immunity that is protective for poultry against HPAI, the results of the phase-3 trial and the absence of HPAI in any of the fully vaccinated farms in 2002-03 in the face of circulating HPAI virus suggests that the levels achieved provide adequate flock immunity.

Commencing vaccination at an older age and farm of origin was significantly associated with poor antibody response. At the commencement of the programme batches of chickens up to 55 days of age were vaccinated. Experience in Hong Kong generally and in the current vaccination evaluation study was that birds in the 20 to 50-day age range are exposed to and are often susceptible to immunosuppressive effects from diseases such as IBD and ND. Initial vaccination at that age in immunosuppressed birds is likely to give inadequate immune-system priming with consequent poor antibody response from secondary vaccination. The farm-of-origin effect may also relate to the fact that multiple batches of chickens on individual farms, including older birds, were vaccinated at one time. Farmers in Hong Kong have limited experience of vaccinating older birds (there are no broiler breeders in Hong Kong).

Independent evaluation of the ability of the killed H5N2 vaccine to protect against experimental challenge with current Hong Kong H5N1 virus was conducted by

Professor Robert Webster's group at the WHO Reference Laboratory for Avian Influenza in Memphis, Tennessee, USA. They showed that chickens vaccinated at the same ages as for the field vaccination programme and challenged 3 weeks later with another highly pathogenic Hong Kong strain of H5N1 virus ('Z' genotype from the index farm in the February 2002 Hong Kong outbreak) were significantly protected from disease and showed reduced levels of virus excretion compared with unvaccinated controls (R.G. Webster, personal communication).

Although vaccination in the face of an outbreak was tried and on 2 farms was able to protect birds and shut down virus replication, this is not a suitable option in Hong Kong where all farms are in relatively close proximity due to limited land availability. It is very difficult to define and control an epidemiologically sustainable perimeter for infected and dangerous contact areas around which ring vaccination could be used. This would also require vaccination of older birds which may give sub-optimal antibody responses, and there is also potentially a greater chance of selection of variant viruses if virus is replicating rapidly in the presence of partial or incomplete flock immunity than there would be if virus is introduced to a fully vaccinated flock that has had time to develop its immunity.

While the ultimate goal should be to eradicate highly pathogenic avian influenza viruses when they occur, the presence of these viruses in wild birds in the Southern-China region means that this is not possible at present and the risk of infection in Hong Kong is very high. Enhanced biosecurity can reduce this risk but vaccination provides an additional layer of protection and its use is fully justified under these special circumstances.

Concerns have been raised that the use of influenza vaccines in poultry may accelerate antigenic drift of the virus necessitating frequent changes of vaccine composition. In addition, it has been suggested that there may be prolonged but undetected virus shedding in vaccinated chickens, that there will be delays in detecting emerging strains and that vaccination may undermine the push for enhanced biosecurity.

Vaccination elsewhere does not appear to have increased the risk of selection of new strains of virus. In sequential characterization of H5N2 viruses from Mexico, where AI vaccination has been practiced most widely, the Southeast Poultry Research Laboratory, USDA (SEPRL) has not demonstrated any acceleration of 'drift' and the vaccines are still protective (D.E. Swayne, personal communication). In experimental studies at SEPRL, H5 avian-influenza vaccines could protect against H5 viruses isolated from four continents over a 38-year period, despite variation of up to 10.9% in deduced amino-acid sequence of the haemagglutinins (Swayne et al. 1999; 2000).

In fact, evolution and selection of highly pathogenic avian influenza viruses occurs in the absence of vaccination. Low or mildly pathogenic avian influenza viruses in USA (1983-84), Mexico (1994-95) and Italy (1999-2000) have mutated to highly pathogenic influenza viruses without the influence of vaccination (D.E. Swayne, personal communication).

Field experience in a H7 outbreak in turkeys in Utah in 1995 (Halvorson et al. 1997) and on two farms in the 2003 H5N1 outbreak in Hong Kong indicated that new cases of disease stopped in vaccinated flocks and the virus was eliminated. There is no evidence to suggest that vaccination led to prolonged undetected shedding of virus. As vaccinated birds exposed to H5N1 viruses shed far less virus than their non-vaccinated counterparts it is likely that infection in properly vaccinated flocks will be self-limiting.

The monitoring programme instigated on vaccinated chicken farms in Hong Kong involves 60 individually identified unvaccinated sentinel chickens within each batch of chickens. The sentinels are monitored serologically, clinically and if required virologically over the 90 to 100-day production cycle for evidence of H5 avian influenza virus infection. This is relatively easy to achieve with the cage-rearing system used in Hong Kong. 'DIVA' (differentiating infected from vaccinated animals) serological testing has been implemented elsewhere as an aid to detecting avian-influenza infection in vaccinated flocks (Capua et al. 2003b). However, DIVA testing is likely to be less sensitive than sentinel-bird monitoring when dealing with a highly pathogenic influenza virus that kills virtually all infected non-immune birds.

The results from this study demonstrate that killed H5N2 influenza-virus vaccine is suitable for inclusion into control programmes for H5N1 avian influenza in Hong Kong. However, vaccination will only be used as part of a package of measures including enhanced biosecurity programmes for farms, wholesale and retail poultry markets, the use of rest days in markets to break cycles of infection and a comprehensive monitoring and surveillance programme for rapid detection of any H5 avian influenza virus incursions. The latter includes dead-bird testing from farms, wholesale and retail markets and a regular programme of viral culturing of cage swabs from farms and retail markets. The farm monitoring also includes antibody testing to ensure vaccinated flocks maintain adequate H5 antibody levels.

Acknowledgments

We thank the staff of the Avian Influenza Serology, Avian Virology, Molecular Biology, Histology and Bacteriology laboratories at Tai Lung Veterinary Laboratory for their excellent technical support and field staff of Livestock Farm Division for the supervision of the on-farm vaccination, bird identification, clinical inspection of vaccinated and sentinel birds, blood and virology sample collection, which were all conducted in a highly competent manner. The assistance provided in the data analysis of the Phase-1 vaccination trial by Professor T.B. Farver, Department of Population Health and Reproduction, University of California, Davis is gratefully acknowledged.

References

Alexander, D.J., 2000. Highly pathogenic avian influenza. *In: OIE manual of standards for diagnostic tests and vaccines*. 4th edn. World Organisation for Animal Health OIE, Paris, 212-220.

Capua, I., Marangon, S., Dalla Pozza, M., et al., 2003a. Avian influenza in Italy 1997-2001. *Avian Diseases,* 47 (Suppl. S), 839-843.

Capua, I., Terregino, C., Cattoli, G., et al., 2003b. Development of a DIVA (Differentiating Infected from Vaccinated Animals) strategy using a vaccine containing a heterologous neuraminidase for the control of avian influenza. *Avian Pathology,* 32 (1), 47-55.

Collins, R.A., Ko, L.S., So, K.L., et al., 2002. Detection of highly pathogenic and low pathogenic avian influenza subtype H5 (Eurasian lineage) using NASBA. *Journal of Virological Methods,* 103 (2), 213-225.

Halvorson, D.A., 2002. The control of H5 or H7 mildly pathogenic avian influenza: a role for inactivated vaccine. *Avian Pathology,* 31 (1), 5-12.

Halvorson, D.A., Frame, D.D., Friendshuh, A.J., et al., 1997. Outbreaks of low pathogenicity avian influenza in USA. *In:* Slemons, R.D. ed. *Proceedings of the 4th international symposium on avian influenza, held May 29-31, 1997.* US Animal Health Association, Georgia Center for Continuing Education, The University of Georgia, Athens, 36-46.

Naeem, K., 1997. The avian influenza H7N3 outbreak in South Central Asia. *In:* Slemons, R.D. ed. *Proceedings of the 4th international symposium on avian influenza, held May 29-31, 1997.* US Animal Health Association, Georgia Center for Continuing Education, The University of Georgia, Athens, 31-35.

Shortridge, K.F., 1999. Poultry and the influenza H5N1 outbreak in Hong Kong, 1997: abridged chronology and virus isolation. *Vaccine,* 17 (Suppl. 1), S26-S29.

Shortridge, K.F., Gao, P., Guan, Y., et al., 2000. Interspecies transmission of influenza viruses: H5N1 virus and a Hong Kong SAR perspective. *Veterinary Microbiology,* 74 (1/2), 141-147.

Spackman, E., Senne, D.A., Myers, T.J., et al., 2002. Development of a real-time reverse transcriptase PCR assay for type A influenza virus and the avian H5 and H7 hemagglutinin subtypes. *Journal of Clinical Microbiology,* 40 (9), 3256-3260.

Studies with inactivated influenza virus vaccines, 2003. *Weekly Epidemiological Record,* 78 (9), 60. [http://www.who.int/entity/wer/2003/en/wer7809.pdf]

Swayne, D.E., Beck, J.R., Garcia, M., et al., 1999. Influence of virus strain and antigen mass on efficacy of H5 avian influenza inactivated vaccines. *Avian Pathology,* 28 (3), 245-255.

Swayne, D.E., Beck, J.R., Perdue, M.L., et al., 2001. Efficacy of vaccines in chickens against highly pathogenic Hong Kong H5N1 avian influenza. *Avian Diseases,* 45 (2), 355-365.

Swayne, D.E., Garcia, M., Beck, J.R., et al., 2000. Protection against diverse highly pathogenic H5 avian influenza viruses in chickens immunized with a recombinant fowlpox vaccine containing an H5 avian influenza hemagglutinin gene insert. *Vaccine,* 18 (11/12), 1088-1095.

Villarreal, C.L. and Flores, A.O., 1997. The Mexican avian influenza H5N2 outbreak. *In:* Slemons, R.D. ed. *Proceedings of the 4th international symposium on avian influenza, held May 29-31, 1997.* US Animal Health Association, Georgia Center for Continuing Education, The University of Georgia, Athens, 18-22.

Wong, H., 2003. *Evaluation of a chicken vaccination programme with an inactivated H5N2 avian influenza vaccine to prevent outbreaks of highly pathogenic H5N1 avian influenza in Hong Kong.* Thesis University of California, Davis.

10

Vaccination of poultry against avian influenza: epidemiological rules of thumb and experimental quantification of the effectiveness of vaccination

M. van Boven[1], J. van der Goot[2], A.R.W. Elbers[2], G. Koch[2], G. Nodelijk[1], M.C.M. de Jong[1], T.S. de Vries[3], A. Bouma[4] and J.A. Stegeman[4]

Abstract

In The Netherlands a large outbreak of highly pathogenic avian influenza in poultry occurred in 2003. The outbreak has had devastating consequences, from both economic and animal-health perspective. Vaccination of poultry offers a potentially attractive measure to control and prevent outbreaks of highly pathogenic avian influenza. In this paper we discuss, from an epidemiological perspective, the values and limitations of vaccination as a control measure during an outbreak and as a preventive measure in an area at risk. In particular, we will discuss (*i*) the epidemiological prerequisites that have to be met for a vaccine and vaccination campaign to be effective, and (*ii*) experimental data that have helped quantifying the effect of vaccination on the reduction of transmission levels. We also discuss (*iii*) how the theoretical insights and experimental results have assisted the Dutch authorities to decide on whether or not to implement vaccination as a control measure.

Keywords: outbreak; control; herd immunity; epidemic model; statistical analysis

Introduction

Low-pathogenicity avian influenza A (LPAI) viruses in poultry of the H5 and H7 subtypes are noted for their ability to transform into highly pathogenic counterparts (HPAI). Outbreaks of HPAI virus in poultry usually result in considerable damage, from both an economic and an animal-health perspective. At least 20 outbreaks of avian influenza have been recorded in poultry since 1959 (Alexander 2000; Alexander et al. 2000).

In this paper we will discuss the value of vaccination as a control measure during an outbreak and as a preventive measure in an area at risk. In particular, in the next section (*Characteristics of an effective vaccination campaign*) we will discuss the conditions that have to be met in order for a vaccination campaign to reduce transmission levels to such an extent that it can halt or prevent epidemic outbreaks of HPAI. In the following section (*Experimental quantification of transmission*) we will

---

[1] Animal Sciences Group, Wageningen University and Research Centre, PO Box 65, 8200 AB Lelystad, The Netherlands. E-mail: michiel.vanboven@wur.nl
[2] Central Institute for Animal Disease Control, Wageningen University and Research Centre, PO Box 2004, 8203 AA Lelystad, The Netherlands
[3] Animal Health Service, PO Box 9, 7400 AA Deventer, The Netherlands
[4] Faculty of Veterinary Medicine, Utrecht University, Yalelaan 7, 3584 CL Utrecht, The Netherlands

R. S. Schrijver and G. Koch (eds.), Avian Influenza, 85–92.
© 2005 *Springer. Printed in the Netherlands.*

then examine the experimental evidence that has accumulated to quantify the effect of vaccination of the transmission dynamics of HPAI virus. In particular, we will show how the so-called basic reproduction number can be quantified to judge the effectiveness of a vaccine and vaccination campaign. Finally, in the last section (*Recommendations given during the Dutch outbreak*) we will indicate how the theoretical rules of thumb and experimental data may have helped policymakers to decide on whether or not to implement vaccination as a control measure.

Characteristics of an effective vaccination campaign

At first sight, vaccination seems an attractive option to prevent or control outbreaks of HPAI in poultry. Indeed, if a vaccine were available that could provide immediate and complete protection against infection, and if such a vaccine could be quickly administered to all animals in a certain area, a developing epidemic could certainly be halted by vaccination. It is, however, unlikely that a perfect vaccine will be available in the near future. Furthermore, there are considerable practical difficulties that need to be overcome before vaccination becomes a viable option (How can millions of animals be vaccinated in a fairly short time span? Which components should be included in the vaccine?).

In the next subsections we will lay bare, in the idealized context of a large unstructured population of hosts, what a vaccine and vaccination campaign should be able to achieve in order to control or prevent outbreaks of HPAI virus.

The reproduction number

Vaccination of poultry against infection with HPAI virus can have a number of objectives: (*i*) to reduce morbidity and/or mortality after infection with HPAI virus; (*ii*) to reduce transmission of HPAI virus within and between farms; and (*iii*) to reduce transmission within and between farms to an extent that it is sufficient to halt an epidemic.

From an epidemiological point of view, objective (*i*) has little value as it may do nothing to prevent the continuing spread of the virus from animal to animal, and from farm to farm. Objective (*ii*) may have some merits from an epidemiological perspective, as it may slow down the spread of the virus and may ultimately result in fewer animals and farms being affected. However, objective (*ii*) by no means guarantees that a vaccination campaign can prevent or effectively contain a major outbreak of HPAI virus. To prevent a large outbreak of HPAI or to fight off an outbreak once it has started, it is necessary to reduce transmission to such a level that, on average, each infected animal (or farm) infects less than *1* other animal (or farm). This quantity is commonly referred to as the (basic) reproduction number, or R.

Let us consider a simple model that contains several important features of HPAI-virus transmission in poultry. In the following we denote by g_1 and g_2 the susceptibility to infection of animals that are vaccinated and unvaccinated, respectively. Likewise, we denote by f_1 and f_2 the infectiousness of infected animals that are vaccinated and unvaccinated, respectively. Furthermore, we denote the relative frequency of infected vaccinated and unvaccinated animals by I_1 and I_2, respectively. Using the above notation the rates at which vaccinated animals are infected by vaccinated and unvaccinated animals are given by $g_1 f_1 I_1$ and $g_1 f_2 I_2$, respectively. Likewise, the rates at which unvaccinated animals are infected by vaccinated and unvaccinated animals are given by $g_2 f_1 I_1$ and $g_2 f_2 I_2$, respectively.

Finally, we denote by ρ_1 and ρ_2 the rates at which vaccinated and unvaccinated animals recover from the infection, and by μ_1 and μ_2 the rates at which animals die from the infection, respectively.

It is now standard practice to derive the reproduction number from the above model formulation (e.g. Diekmann and Heesterbeek 2000). The reproduction number R can be expressed in terms of the parameters as follows:

$$R = \frac{f_1 g_1}{\rho_1 + \mu_1} p + \frac{f_2 g_2}{\rho_2 + \mu_2}(1 - p)$$

$$= R_{vac} p + R_{unvac}(1 - p) , \tag{1}$$

where p denotes the fraction of the population that is vaccinated, $R_{unvac} = \dfrac{f_2 g_2}{\rho_2 + \mu_2}$ denotes the reproduction number in a population of susceptible animals, and $R_{vac} = \dfrac{f_1 g_1}{\rho_1 + \mu_1}$ represents the reproduction number in a population that consists entirely of vaccinated animals.

Minor versus major outbreaks

The reproduction number can be used to estimate the probability that an outbreak starting with one or a few infected animals will not by chance come to a standstill (a *minor outbreak*). In fact, under certain assumptions (Diekmann and Heesterbeek 2000), the probability f of a so-called *major outbreak* that affects a non-negligible fraction of the animals can be determined from the equation

$$1 - f = e^{-Rf} . \tag{2}$$

There is considerable scope for extension of the above result, but we will not dwell on this issue here.

Herd immunity and the critical vaccination effort

The fundamental property of a successful vaccination campaign is that it should provide *herd immunity* (Anderson and May 1991). The concept of herd immunity implies that vaccination does not only have a direct protective effect for those animals that are vaccinated, but also decreases the probability of infection for animals that are not vaccinated. In fact, in a population with herd immunity the transmission opportunities are decreased to such an extent that no long-lasting infection chain can be sustained.

For the model introduced above, the *critical vaccination fraction* at which herd immunity is achieved is found by putting $R=1$ in Equation (1), and solving the equation for $p=p_c$. The critical vaccination fraction is given by

$$p_c = \frac{1 - R_{unvac}}{R_{vac} - R_{unvac}} . \tag{3}$$

Notice that eradication of the pathogen is not possible whenever $R_{vac} > 1$, i.e. when vaccination does not prevent the continued spread of the virus amongst vaccinated animals (of course assuming $R_{unvac} > R_{vac}$). In case that vaccination completely blocks transmission to or from vaccinated animals $R_{vac} = 0$, and Equation (3) reduces to the familiar $p_c = 1 - \dfrac{1}{R_{unvac}}$ (Anderson and May 1991).

Experimental quantification of transmission

How can the above theoretical quantities be quantified, and how can the results be applied to real-world situations? Typically, it will be very difficult to obtain reliable indicators of pathogen and vaccine characteristics during the course of an outbreak, and one has to resort to *transmission experiments* to quantify the characteristics of certain pathogen–vaccine combinations (infectious period, virulence, infectiousness) (see De Jong and Kimman (1994) for an early application of these methods). In a transmission experiment a number of animals that are inoculated with the virus are put into a stable with a number of susceptible contact animals. The ensuing infection chain is monitored on a regular basis by taking swabs from the cloaca and trachea that are subject to virus-isolation techniques and PCR, and by collecting blood samples to determine antibody titres.

For practical reasons, the number of animals used in a transmission study is usually rather small (in the order of 10 to 20). Therefore the above theoretical results for large populations need to be adapted to small population sizes where several chance effects play an important role. In the next two subsections we will therefore shortly introduce the methods on which the statistical analysis of the experimental epidemics is based. The account in the next two subsections is based largely on the paper by Van der Goot et al. (2003a). In the last subsection we will give an overview of the experiments that have been carried out (see Van der Goot et al. 2003a; 2003b; in prep.-b; in prep.-a, for details).

Final size analysis

In first instance, the analysis of the experiments is based on the final size of the experiments, i.e. the number of contact animals that have been infected when the infection chain has ended. The final sizes are used to obtain estimates of the (basic) reproduction number, i.e. the number of infections that would be caused by a single infected animal in a large population of susceptible animals. A forte of final-size methods is that they are robust (e.g., inclusion of a latent period does not alter the results) and that different assumptions on the distribution of the infectious period are easily incorporated.

The methods are based on maximum-likelihood estimation (MLE). MLE can be used because final-size distributions can be determined under a wide range of assumptions. We refer to Ball (1995) for a fairly accessible introduction in final-size methods.

Generalized Linear Model

Final-size methods are flexible but do not make full use of the available information. To take the time course of the experimental epidemics into account, we estimate the transmission parameter β of the stochastic SIR model by means of a Generalized Linear Model (GLM). We refer to Becker (1989) for an introduction of GLMs in the context of epidemic models.

To apply a GLM the data of the experiments are first rendered into the format (S, i, C). Here S is the number of susceptible animals in a certain time period, i is the prevalence of infection (i.e. the average number of infectious animals divided by the total number of animals), and C represents the number of new infections that have appeared at the end of the time period. By standard reasoning we assume that the number of cases C arising in a day is binomially distributed with parameter $p_{inf} = 1 - e^{-\beta i}$ (the probability of infection) and binomial totals S:

$$C \sim Bin(S, 1 - e^{-\beta i}).\tag{4}$$

Experiments

Several experiments have been carried out with LPAI and HPAI viruses of the H5 and H7 subtypes. Table 1 gives an overview of the experiments. The first three sets of experiments involved LPAI and HPAI strains of the H5N2 subtype that were isolated during a large outbreak of HPAI in Pennsylvania in 1983. The results of these experiments are published in Van der Goot et al. (2003a; 2003b).

The next two sets of experiments were done using LPAI and HPAI strains of the H7N1 subtype, isolated during the outbreak of avian influenza in Italy in 1999. These experiments form part of the EU project AVIFLU on avian influenza. The last three sets of experiments were carried out using an HPAI strain of the H7N7 subtype that was isolated in The Netherlands during the outbreak of 2003. These experiments were also carried out as part of the EU project AVIFLU.

The analysis of the experiments with the H5N2 strains and the implications of the results are described in Van der Goot et al. (2003a; 2003b). The results of the experiments with H7N1 and H7N7 will be published by Van der Goot (in prep.-b; in prep.-a).

subtype and strain	control measure	no. replicates	remarks
A/chicken/Pennsylvania/83 H5N2 (LPAI)	none	4	Van der Goot et al. (2003a; 2003b); financed by Dutch government
A/chicken/Pennsylvania/83 H5N2 (HPAI)	none	2	Van der Goot et al. (2003a; 2003b); financed by Dutch government
A/chicken/Pennsylvania/83 H5N2 (HPAI)	previous infection with H5N2 (LPAI)	2	Van der Goot et al, (2003a; 2003b); financed by Dutch government
A/chicken/Italy/99 H7N1 (LPAI)	none	2	part of EU project AVIFLU/ financed by Dutch government
A/chicken/Italy/99 H7N1 (HPAI)	none	2	Van der Goot et al. (in prep.-a); part of EU project AVIFLU/ financed by Dutch government
A/chicken/Netherlands/03 H7N7 (HPAI)	none	2	Van der Goot (in prep.-b; -a); part of EU project AVIFLU/ financed by Dutch government
A/chicken/Netherlands/03 H7N7 (HPAI)	heterologous vaccination (H7N1)	4	Van der Goot et al. (in prep.-b); part of EU project AVIFLU/ financed by Dutch government
A/chicken/Netherlands/03 H7N7 (HPAI)	heterologous vaccination (H7N3)	4	Van der Goot et al. (in prep.-b); part of EU project AVIFLU/ financed by Dutch government

Table 1. Transmission experiments carried out with avian influenza A viruses of the H5 and H7 subtypes

Recommendations given during the Dutch outbreak

The above theoretical rules of thumb and experimental evidence on the transmission characteristics formed the basis on which the Dutch authorities were advised on vaccination as a pre-emptive measure to prevent outbreaks of HPAI or as a control measure during an outbreak of HPAI (Van Boven et al. 2003). Shortly, the questions posed were the following:

1. Could **vaccination of a ring** around an infected farm be an effective control measure? If so, what would be the ideal 'vaccination radius' and how many farms would have to be vaccinated?
2. Would **vaccination of a compartment** (e.g., a province) once an infected farm has been detected in a compartment be effective to prevent the spread of the virus to other compartments? How large should the compartments be (in terms of area and number of farms)?
3. Is **preventive vaccination** of an area at risk a viable option? In particular, how does vaccine efficacy depend on vaccine composition? What is the maximal allowable rate of (primary) vaccine failure? Is it necessary to vaccinate all farms and animals in an area?
4. How should **repopulation** of an area in which HPAI has circulated before be carried out? Which farms should be repopulated first, and at what density? Should a repopulation programme be accompanied by a vaccination and/or surveillance programme?
5. What properties should a **surveillance programme** have in order to make sure that there is no transmission of influenza virus and that introductions of virus are noticed sufficiently quickly?

It is clear that, given the limited amount of information on the transmission dynamics of HPAI between flocks and given the very limited experience with vaccination applied as a systematic control measure (see Capua and Marangon (2003) for an exception), no definitive answers can be given to the above questions. Therefore, our advice was based on (1) the general epidemiological principles mentioned above, and (2) the developing experience with transmission experiments in a small population of poultry. The text below is translated from the Dutch report of Van Boven et al. (2003). We refer to Van der Goot et al. (in prep.-b) for an up-to-date account of transmission experiments in vaccinated poultry.

1. With a view to the very fast dynamics of AI in poultry flocks and the fact that it takes some weeks before a vaccination programme offers protection, **ring vaccination** is useless from an epidemiological point of view, unless the ring is quite large (a radius of >50 km). Moreover, in practice there will be an extra delay because it takes a while before all the farms in the area have been vaccinated, which only enlarges the area and the number of farms to be vaccinated.
2. It is extremely doubtful whether **a vaccination campaign in a large area/compartment** is at all useful. As it takes a relatively long time before the vaccine offers effective protection (2 to 4 weeks), and extra time to vaccinate all the farms (>1 week), it is quite unlikely that a vaccination programme that is started at the moment when an infection has been found will be effective.
3. The epidemiological analysis of the outbreaks in the 'Gelderse Vallei' and the province of Limburg indicate that there are **two risk areas in The Netherlands**: the 'Gelderse Vallei' and the area around the town of Weert in Limburg. As the density of poultry farms in these areas is very high, a chain reaction of new infections may arise after the introduction of the virus. The rest of The Netherlands does not seem to be an epidemiological risk area: an introduction of the virus is likely to result in few additional infected farms if the virus is detected in time and if effective measures are taken fast (closing down the farm and strict transport restrictions in an area around the focus of infection).

4. **Preventive vaccination** of an area/compartment before an outbreak has been spotted could be useful from an epidemiological point of view. Whether such a preventive vaccination is actually effective depends on 1) the effectivity of the vaccine in decreasing or blocking the transmission of avian influenza virus to and from vaccinated farms, and 2) the fraction of farms in the area that is vaccinated. General epidemiological principles say that the number of infected farms during an outbreak can only be kept small if the between-farm reproduction number R_h is brought below 1. The estimates of R_h during the outbreaks varied from 4 to 6. This means that if the vaccine blocks farm-to-farm transmission completely at least 75 to 84 % of the farms must be vaccinated. If the vaccine does not completely block the transmission between farms, then a higher percentage of the farms must be vaccinated.

5. **Repopulation** of farms in previously poultry-dense areas constitutes a high risk for the re-occurrence of the virus in poultry. In Italy repopulation programmes after a primary outbreak have repeatedly led to new outbreaks. Should this happen in The Netherlands in poultry-dense areas, then this could lead to a chain reaction of new infections. Therefore, from an epidemiological point of view it is advisable to re-populate the 'Gelderse Vallei' and Limburg in phases. As long as the density of farms remains lower than a certain critical density a reintroduction on a farm will in all probability not lead to an explosive wave of newly infected farms. Vaccination in the case of repopulation could be a useful additional measure besides better surveillance, improved hygiene and transport restrictions because it can reduce the effective density of farms that are at risk. To detect the possibly still present virus as soon as possible and with minimal cost farms that have been infected during the outbreak must be re-populated first.

6. Detecting new introductions of avian influenza virus fast is a crucial part of fighting the virus effectively. It is therefore advisable to launch a large-scale **surveillance programme** when repopulating areas at risk. Also, a surveillance programme can help monitoring the areas at risk for introduction along the border.

7. If it is decided to carry out a large-scale surveillance programme an effective test (i.e. a test with sufficiently high sensitivity) should be available that can distinguish between animals that have been vaccinated and animals that have suffered a natural infection. Such a test that can distinguish infection from vaccination plays an important part in a surveillance programme in a vaccinated area to detect infected farms as quickly as possible.

8. Since vaccination probably does not completely block transmission of the virus a vaccination programme can be useful only as an additional measure. The **real danger** is that because of vaccination a misplaced sense of security is created and that the necessary basic measures such as hygiene, transport restrictions and surveillance are no longer observed ("after all, we are vaccinating"). For that reason it is of the greatest importance to make sure that any possible vaccination campaigns are always accompanied by the necessary flow of information.

Acknowledgments

Miss B. Prins is gratefully acknowledged for linguistic advice and translation of the Dutch text of the last section.

References

Alexander, D.J., 2000. A review of avian influenza in different bird species. *Veterinary Microbiology,* 74 (1/2), 3-13.

Alexander, D.J., Meulemans, G., Kaleta, E., et al., 2000. *The definition of avian influenza and the use of vaccination against avian influenza.* EU report of the Scientific Committee on Animal Health and Animal Welfare. [http://europa.eu.int/comm/food/fs/sc/scah/out45_en.pdf]

Anderson, R.M. and May, R.M., 1991. *Infectious diseases of humans: dynamics and control.* Oxford University Press, Oxford.

Ball, F., 1995. Coupling methods in epidemic theory. *In:* Mollison, D. ed. *Epidemic models: their structure and relation to data.* Cambridge University Press, Cambridge, 35-53.

Becker, N.G., 1989. *Analysis of infectious disease data.* Chapman and Hall, London.

Capua, I. and Marangon, S., 2003. Vaccination policy applied for the control of avian influenza in Italy. *In:* Roth, J.A. ed. *Vaccines for OIE list A and emerging animal diseases: international symposium, Ames, Iowa, USA, 16-18 September, 2002: proceedings.* Karger, Basel, 213-219. Developments in Biologicals no. 114.

De Jong, M.C.M. and Kimman, T.G., 1994. Experimental quantification of vaccine-induced reduction in virus transmission. *Vaccine,* 12 (8), 761-766.

Diekmann, O. and Heesterbeek, J.A.P., 2000. *Mathematical epidemiology of infectious diseases: model building, analysis and interpretation.* John Wiley, Chichester.

Van Boven, M., Boender, G., Elbers, A., et al., 2003. *Epidemiologische consequenties van vaccinatie.* Rapport voor het Ministerie van Landbouw, Natuur en Voedselkwaliteit. [http://www9.minlnv.nl/pls/portal30/docs/FOLDER/MINL NV/LNV/STAF/STAF_DV/KAMERCORRESPONDENTIE/2003/BIJLAGE N/PAR03244A.PDF]

Van der Goot, J., Koch, G., De Jong, M.C.M., et al., in prep.-a. Comparison of the transmission characteristics of highly pathogenic avian influenza viruses in poultry (H5N1, H5N2, H7N1 & H7N7).

Van der Goot, J., Koch, G., De Jong, M.C.M., et al., in prep.-b. Quantification of the transmission characteristics of highly pathogenic avian influenza A virus (H7N7) in vaccinated and unvaccinated chickens.

Van der Goot, J.A., De Jong, M.C.M., Koch, G., et al., 2003a. Comparison of the transmission characteristics of low and high pathogenicity avian influenza A virus (H5N2). *Epidemiology and Infection,* 131 (2), 1003-1013.

Van der Goot, J.A., Koch, G., De Jong, M.C.M., et al., 2003b. Transmission dynamics of low- and high-pathogenicity A/Chicken/Pennsylvania/83 avian influenza viruses. *Avian Diseases,* 47 (Special issue), 939-941.

11

The development of avian influenza vaccines for emergency use

T.R. Mickle[#], D.E. Swayne[##], and N. Pritchard[###]

Abstract

Costly outbreaks of mildly and highly pathogenic avian influenza (AI) have occurred in the commercial poultry industry in Europe and the United States in the past two years. The current approach is to control the disease by depopulation of infected flocks followed by cleaning and disinfection of the premises. The cost of eradication of influenza and the payments to the poultry producer continue to increase. The cost of the AI eradication in the Netherlands and the United States was more than 500 million USD. The use of vaccines to control AI is gaining acceptance by veterinary health agencies as a tool in eradication programmes. The choice of vaccines available includes purified subunit vaccines, genetically modified vaccines and the traditional whole-virus inactivated vaccines. The use of inactivated vaccines has been used successfully in many countries to stop the spread of avian influenza in the poultry industry. The fowlpox-vectored vaccine TROVAC AI H5™ has been used to vaccinate broiler chickens in Mexico for five years. The preparation of a supply of vaccine in advance of a disease outbreak has been used in the human health sector. A vaccine bank was created at Merial for foot-and-mouth disease more than 10 years ago. The idea of developing a vaccine bank for avian influenza is being discussed in the United States and in the European Union. Before a strategic plan for AI vaccines can be implemented, many questions about the AI strains needed, the amount of vaccine, the formulation, the priority of vaccination in the poultry industry and the cost to produce and maintain stored antigens or vaccines need to be addressed.

Keywords: avian influenza; vector vaccine; fowlpox; antigen or vaccine bank; TROVAC AI H5™; LPAI; HPAI

Introduction

For more than 30 years inactivated whole-virus avian influenza vaccines have been the only product available to control the spread of the disease from infected to susceptible birds. However, due to an international agreement, animal-health regulatory agencies relied on the destruction and removal of infected birds from susceptible ones. The number of outbreaks in the poultry-producing countries of the world has increased during the last 10-15 years. The United States had low-

---

[#] Merial Limited Avian Global Enterprise, 3239 Satellite Boulevard, Duluth, Georgia 30096, USA. E-mail: tom.mickle@merial.com
[##] USDA-ARS, Southeastern Poultry Research Lab., 934 College Station Road, Athens, Georgia 30605, USA
[###] Merial Select, Inc. P.O. Drawer 2497, Gainesville, Georgia 30501 USA

R. S. Schrijver and G. Koch (eds.), Avian Influenza, 93–100.
© 2005 *Springer. Printed in the Netherlands.*

pathogenic AI infections in commercial layer complexes in the states of Connecticut and Texas. Turkeys and chickens were infected in the outbreak of low-pathogenic AI type H7 in the state of Virginia. In Europe the disease of both low and high pathotypes affected the commercial poultry industry of The Netherlands and Italy. The cost of eradicating the infected flocks runs in the millions of dollars. The magnitude of the outbreaks has increased over the years due to the increasing concentration of poultry production around the world.

Inactivated whole-virus AI vaccines

The role of and a justification for the use of inactivated AI vaccines for the control of mildly and highly pathogenic avian influenza has been described (Halvorson 2002). The state of Minnesota has used monovalent inactivated vaccines of several haemagglutinin subtypes in the turkey industry to avoid the destruction of infected flocks. An inactivated oil-emulsion vaccine was used to immunize 4 million layers in the state of Connecticut this year. Hong Kong started an experimental programme of vaccination using an inactivated H5 vaccine produced in Mexico. The experience with the use of inactivated H7 AI vaccine in Italy will be presented in this volume.

Inactivated AI vaccines are relatively simple to prepare and provide high levels of immunity when properly manufactured and administered. Merial produces five different inactivated AI vaccines (Table 1). Most of the AI vaccines have been requested by governments in the Middle East for endemic situations in the poultry industry. Merial has manufactured an inactivated H7N1 oil-emulsion vaccine for use in Italy.

One disadvantage of inactivated vaccines is that every bird must be inoculated individually using a syringe. The vaccination of birds at the farm is difficult and stressful on the birds. The time required to inject each bird depends on the age of the bird being vaccinated, the volume and viscosity of the vaccine and the number of people that are available to catch and vaccinate the flock. The cost of labour and the threat of spreading the disease with vaccination crews is a major concern. Attempts at replacing whole-virus vaccines with genetically modified vaccines have not occurred as quickly as expected.

Table 1. Inactivated AI vaccines

Inactivated AI vaccine name	Avian influenza type
GALLIMUNE FLU™	H7N1
BIOFLU/BIOENFLU	H6N2 & H9N2
FLUVAC	H7N3
GALLIMUNE FLU™	H9N2
GALLIMUNE 208™	H9N2 + Newcastle

New technologies for vaccines

Biotechnology has given us alternatives to the whole-virus vaccine and the potential for developing quantities of antigen or finished vaccines that can be stored for use in the event of a future outbreak. The feasibility of several different types of genetically engineered avian influenza vaccines has been demonstrated. However, only the fowlpox-virus-vectored vaccine has been used continuously in an AI-control effort.

Protein Sciences, Inc. developed a subunit vaccine of recombinant HA glycoprotein utilizing a baculovirus–insect cell expression system (Wilkinson 1997). The experimental vaccine with an adjuvant provided a good serological response and protected vaccinated SPF chickens against challenge with a lethal strain of HPAI. The rHA without the addition of an adjuvant provided only partial protection. The subunit vaccine also requires individual injection of each bird. So far this approach has not been commercialized for veterinary use.

Kodihalli and Webster (1997) reported on the efficacy of DNA vaccines delivered with a gene gun at the Fourth International Symposium on Avian Influenza. Delivery of a single injection of DNA encoding the influenza HA provided immunity for the life of the chicken. The method for DNA injection depends on the use of the gold beads to carry the antigen into the tissue of the animal. The cost of the sophisticated injection system and the precision required to inject a chicken properly with DNA has prevented the mass application of the technology in the poultry industry.

The feasibility of using fowlpox as a vector to express the HA gene of AI was described in 1988 by researchers at Virogenetics and the New York Department of Health. The haemagglutinin gene from subtype H5 was successfully expressed in fowlpox (Taylor et al. 1988). The efficacy of a recombinant fowlpox–AI vaccine against a virulent H5N2 virus was decribed by Beard, Schnitzlein and Tripathy (1991).

Merial began registration of the GMO vaccine TROVAC AI H5 after a request from the US poultry industry in 1995. There was concern that the highly pathogenic AI H5 virus circulating in Mexico would enter the US. The vaccine uses the fowlpox virus from Merial's product named Diftosec, which has been used in Europe for many years. The pox virus was further modified with specific deletions in the genome to ensure safety and genetic stability. The HA gene from A/Turkey/Ireland/83 strain was inserted into a non-essential location in the pox-virus genome resulting in a virus that protect poultry from two infectious diseases, fowlpox and AI.

The vaccine provides excellent protection against a wide range of highly pathogenic subtype-H5 viruses (Swayne et al. 2000). It stimulated an antibody titre faster than inactivated oil-emulsion vaccines and it prevented mortality after exposure to lethal AI viruses for a minimum of 20 weeks. The GMO vaccine was designed to have a differential diagnostic test. The pox vector expresses only one gene, HA. When serum samples from vaccinated chickens are tested using the agar-gel precipitin test or with a commercial ELISA kit the samples will be negative compared to the positive control sample or infected birds. It is the nucleoprotein that is responsible for the positive reaction measured by the ELISA or AGP. The AGP is an inexpensive and simple test to use, however the results may require 24 to 48 hours to confirm. It has been reported that the pox-vectored vaccine was unable to stimulate measurable HI titres.

A recent experiment showed the critical importance of the choice of antigen used in the AI HI test. When the standard A/Turkey/Wisconsin/68 antigen was used in the HI test most of the serum sample titres were measured as 0 or low. However, when the homologous A/Turkey/Ireland/83 was substituted the measurable antibody titres increased. Chickens vaccinated subcutaneously at the day of hatch maintained a HI titre > 1:100 nine weeks after injection.

The TROVAC AI H5™ vaccine can be administered to birds at one-day age or older by wing-web stab or subcutaneous injection. The majority of the doses have been administered in the hatchery using automated injection machines vaccinating chicks with Marek's vaccine. There is a misconception that maternal antibody against the pox vector interferes with the development of immunity. In Mexico the vaccination programme for broiler breeders includes two vaccinations with live fowlpox vaccine. Millions of the progeny are successfully vaccinated every week with

TROVAC AI H5™. More than 700 million doses of fowlpox vectored AI H5 vaccine have been sold in Mexico since registration in 1998.

Previous studies indicated that TROVAC AI H5™ would not provide a consistent immunity when birds were previously vaccinated with live fowlpox vaccine or exposed to wildtype pox vaccine (Swayne, Beck and Kinney 2000). TROVAC AI H5 ™ did not prevent morbidity or provide the necessary 90% efficacy required for licensing when the birds were first immunized with fowlpox vaccine. The addition of an adjuvant may overcome this problem. SPF chickens vaccinated at one day of age with a live fowlpox vaccine and vaccinated twice with TROVAC AI H5™ vaccine and an adjuvant protected 85% of the birds against morbidity and mortality after a challenge with a highly pathogenic AI virus. Although this vaccination programme would not be feasible for broiler chickens it may be possible to design a vaccination programme using two injections of TROVAC AI H5™ for replacement pullets.

Merial has tried to develop a genetically engineered fowlpox-vectored virus containing an H7 gene. A construct identified as vFP1549 has been available for several years. The registration and development programme was delayed because of lack of interest in allowing GMO vaccines as part of an AI-control policy. The efficacy of the construct was proven using a highly pathogenic virus A/Chk/Pakistan/1369-CR2/95.

When the first outbreak of H7N1 was reported in Italy the efficacy of the construct vFP1549 was evaluated using the Italian virus. It was surprising to learn that the fowlpox-vectored AI vaccine failed to protect against the Italian virus A/Turkey/Italy/4580/99 (H7N1). An experiment was done to determine if the route of inoculation or the concentration of the construct would have an effect on the efficacy of the vaccine. The efficacy was slightly better when the chicks were inoculated by subcutaneous injection. However, at concentrations of $10^{2.5}$ $TCID_{50}$ or less the construct did not provide satisfactory immunity.

AI vaccine bank

The Southeastern Poultry and Egg Association formed a working group in 1995 to examine some major questions and develop an action plan for the occurrence of highly pathogenic AI in the United States. One of the recommendations from the task force was to allow for the manufacture of inactivated H5 AI vaccine by qualified vaccine manufacturers. The task force recommended that a financial plan be worked out to pay for the preparation of virus-laden allantoic fluids for storage if an emergency need for AI vaccine occurred. The concept of an antigen bank for avian-influenza vaccines in the United States never became a reality. As the threat of highly pathogenic influenza diminished the necessity for the preparation of a vaccine for emergency use disappeared, too.

The US Department of Agriculture/APHIS, the state veterinarians in poultry-producing states and representatives of the various sectors of the poultry industry began serious discussions concerning the control of both low- and highly pathogenic AI in 2002. The talks began after the outbreak of low-pathogenic AI H7 in the Shenandoah Valley of Virginia. A draft of a series of guidelines was discussed and a resolution for the use of AI vaccines was sent to APHIS by the Transmissible Diseases of Poultry committee at the US Animal Health Association meeting in St. Louis, Missouri in October of 2002. The guidelines included the possibility of using vaccines in conjunction with surveillance and a stamping out to control LPAI and HPAI.

The animal-health community around the world has made provisions for a vaccine bank for foot-and-mouth disease. The concepts that were successfully implemented could be used as a starting point for the creation of an AI bank. There are many factors that must be considered before an action plan can be developed.

Concerns of the biologics industry
The biologics manufacturers will need to accumulate a lot of information to prepare the proper antigens in the bank adequately. Which serotypes are required for storage? The serotypes H5 and H7 must be included because of the reoccurring outbreaks of low- and highly pathogenic viruses. Will the viruses held by the biologics companies be efficacious against the field strain? Some countries will insist that only homologous vaccines be prepared. Who will provide the actual virus for use in the antigen bank?

What type of vaccine should be prepared?
Inactivated oil-emulsion vaccines certainly seem necessary for breeders and layers that are placed in the field. The vectored vaccine could be used to vaccinate susceptible chicks that will continue to be placed in the country. What about the addition of other antigens? In Mexico the vaccine of choice was an inactivated Newcastle Disease-AI vaccine.

Which strains should be included in the vaccine bank?
The bank should include strains of H5 and H7 that are known to cause highly pathogenic avian influenza. Should the bank include MPAI strains? Which neuraminidase strains are needed? When avian influenza is manufactured for a specific country the vaccine is prepared using the strain provided. Would all countries accept a strain of virus with the same serotype even though it may not be the same isolate? Does the poultry industry need vaccines for subtypes H6 and H9?

What type of poultry will be vaccinated?
The target birds will have to be identified. Will the priority of vaccination be given to breeders and layers first? Then vaccine produced for chickens may not be as efficacious for turkeys. Will the formulation for commercial poultry be efficacious to exotic birds confined in zoos? How effective are the inactivated vaccines in ducks or waterfowl? These are important questions that will have to be addressed.

Formulation
Inactivated vaccines manufactured by different companies will have different formulations. The concentration of antigen, adjuvant, emulsion, raw materials will all vary from one source to another. The antigenicity and immunogenicity will vary from one virus to another. When preparing the H7N3 vaccine for Italy it was necessary to conduct nine animal vaccination tests to determine the proper volume of antigen, concentration and volume of the dose necessary for maximum efficacy. The proper formulation is essential if efficacy is going to guaranteed to the to reduce the spread of the disease. If multiple suppliers are going to deposit vaccines in the bank these must be similar enough to use any vaccine without altering the vaccination strategy.

Regulatory requirements
Regulatory-harmonization discussions for animal biologicals between the EU and the US have been going on for several years. Some progress has been made toward

mutual acceptance of products that can be classified as generic vaccines. However, acceptance of a vaccine for a potential zoonotic disease like AI will probably take much longer to resolve. Regulatory harmonization of all the requirements for the production of AI vaccines will be a very important factor. Any vaccine deposited in the bank must be acceptable to all animal-health regulatory agencies around the world. For instance, The Center for Veterinary Biologics in the US requires that antigens of chicken-embryo origin must be prepared in SPF eggs even for inactivated vaccines. This increases the cost of production and quality-control testing compared to Europe where SPF eggs are not required. Any multinational company would prefer to develop a single set of antigens requested by the poultry industry instead of regional antigen supplies based on local or national requirements.

Chain of control

The decision of whether inactivated antigen or final vaccine will dictate who will maintain control If the material stored is antigen then the control will remain with the manufacturer. The biologics company will be responsible for emulsification, filling and release. If final product is stored in the bank then a governmental laboratory could maintain the inventory and assume the responsibility for distributing the product to the end user.

Quality control

The level and exact requirements for quality-control testing will factor into the cost of goods. If the vaccine has to be fully controlled to USDA 9CFR or European Standards the cost of the vaccine will increase. Will the seed material be required to meet all the requirements of a true master-seed concept? The cost of conducting the full range of quality tests required in Europe for a live-virus master seed is approximately 100-150,000 US dollars. The cost will be much less if the requirements are relaxed during an emergency. In the US it has been proposed that in the event of a true epizootic the only requirement for release of final product could be a 24-48 hour sterility test. Will the EU accept the same standards?

Standardization of vaccines

The concentration of haemagglutinin was standardized in an inactivated influenza vaccine using the single-radial-immunodiffusion test produced by Wood et al. (1985). The SRD test provided a simple and reproducible method for standardizing the HA antigen. The SRD data correlated very well with protection against a lethal A/Chicken/Penn./1370/83 virus. A comparison of the efficacy of six inactivated H5N2 vaccines manufactured in Mexico was compared to a standardized oil-emulsion vaccine (Garcia et al. 1998). All of the vaccines prevented signs of AI infection but only one half of the vaccines prevented or reduced viral shedding.

Response time

Although the manufacturing of inactivated oil-emulsion vaccine is a simple technique it is time-consuming and can be labour-intensive. Antigens prepared in embryonated eggs usually require 72 hours of incubation after inoculation with the seed virus. If the AI virus is not well adapted to embryos it can prolong the incubation time. If the inactivation kinetics is already defined a minimum of two more days must be added to the preparation time. Emulsification, filling and bottling, labelling and packaging are additional steps involved for final product. The length of time necessary to prepare inactivated vaccine containing a known antigen could require between 14 and28 days plus the control time. If a vaccine is to be produced with a new field isolate the time to market could be estimated in terms of months.

The fowlpox-vectored vaccine is produced in chicken-embryo-origin fibroblast cells. At Merial we have CEF cell culture available daily. The incubation time required for the fowlpox virus to reach peak virus titre is short. The yields in terms of doses of virus are high. The harvested viral fluids can be frozen and stored or dispensed into glass containers and lyophilized quickly. In a true emergency, millions of doses of TROVAC AI H5™ live vaccine can enter the market in less than seven days including a 24-hour sterility test.

Conclusion

The creation of a vaccine bank is possible but it will require co-operation from the poultry industry, animal-health agencies and the biologics manufacturers. The vaccine companies will strive to provide the products that their customers need to maintain the health of the poultry industry. However, investment in vaccine banks will only occur if there is a certainty that a return on investment for R&D and preparation of the antigens will be received. The preparation of inactivated whole-virus oil-emulsion vaccines requires more lead-time than the fowlpox-vectored vaccine for AI type H5. A best-case estimation for the preparation of an inactivated vaccine for emergency use is 4-6 months. Batches of 20 million doses each of TROVAC AI H5 can be prepared during a five-day work week. Discussions between the poultry scientific community, the poultry companies, regulators and the vaccine companies will need to continue regularly if the AI-vaccine banking system is to become a reality.

Acknowledgments

The authors wish to thank Ms. Joan Beck and Mr. Greg Hanes for their excellent technical assistance.

References

Beard, C.W., Schnitzlein, W.M. and Tripathy, D.N., 1991. Protection of chickens against highly pathogenic avian influenza virus (H5N2) by recombinant fowlpox viruses. *Avian Diseases,* 35 (2), 356-359.

Garcia, A., Johnson, H., Srivastava, D.K., et al., 1998. Efficacy of inactivated H5N2 influenza vaccines against lethal A/Chicken/Queretaro/19/95 infection. *Avian Diseases,* 42 (2), 248-256.

Halvorson, D.A., 2002. The control of H5 or H7 mildly pathogenic avian influenza: a role for inactivated vaccine. *Avian Pathology,* 31 (1), 5-12.

Kodihalli, S. and Webster, R.G., 1997. DNA vaccines for avian influenza: a model for future poultry vaccines? *In:* Swayne, D.E. and Slemons, R.D. eds. *Proceedings of the 4th international symposium on avian influenza, held May 29-31, 1997.* US Animal Health Association, Georgia Center for Continuing Education, The University of Georgia, Athens, 263-280.

Swayne, D.E., Beck, J.R. and Kinney, N., 2000. Failure of a recombinant fowl poxvirus vaccine containing an avian influenza hemagglutinin gene to provide consistent protection against influenza in chickens preimmunized with a fowl pox vaccine. *Avian Diseases,* 44 (1), 132-137.

Swayne, D.E., Garcia, M., Beck, J.R., et al., 2000. Protection against diverse highly pathogenic H5 avian influenza viruses in chickens immunized with a recombinant fowlpox vaccine containing an H5 avian influenza hemagglutinin gene insert. *Vaccine,* 18 (11/12), 1088-1095.

Taylor, J., Weinberg, R., Kawaoka, Y., et al., 1988. Protective immunity against avian influenza induced by a fowlpox virus recombinant. *Vaccine,* 6 (6), 504-508.

Wilkinson, B.E., 1997. Recombinant hemagglutinin subunit vaccine produced in a Baculovirus expression vector system. *In:* Swayne, D.E. and Slemons, R.D. eds. *Proceedings of the 4th international symposium on avian influenza, held May 29-31, 1997.* US Animal Health Association, Georgia Center for Continuing Education, The University of Georgia, Athens, 253-262.

Wood, J.M., Kawaoka, Y., Newberry, L.A., et al., 1985. Standardization of inactivated H5N2 influenza vaccine and efficacy against lethal A/chicken/Pennsylvania/1370/83 infection. *Avian Diseases,* 29 (3), 867-872.

CONTROL MEASURES AND LEGISLATION

12

Should there be a change in the definition of avian influenza for legislative control and trade purposes?

D.J. Alexander[#]

Abstract

The current OIE and EU definitions of avian influenza (AI) to which control measures or trade restrictions should apply were both drafted over 10 years ago. These were aimed at including viruses that were overtly virulent in in-vivo tests and those that had the potential to become virulent. At that time the only virus known to have mutated to virulence was the one responsible for the 1983/84 Pennsylvania epizootic. The mechanism involved has not been seen in other viruses, but the definition set a precedent for statutory control of potentially pathogenic as well as overtly virulent viruses.

Evidence accumulated to date indicates that HPAI viruses arise from LPAI H5 or H7 viruses infecting chickens and turkeys sometime after spread from free-living birds. At present it can only be assumed that all H5 and H7 viruses have this potential and mutation to virulence is a random event. Therefore the longer the presence and greater the spread in poultry the more likely it is that HPAI virus will emerge. The outbreaks in Pennsylvania 1983, Mexico 1994 and Italy 1999 are demonstrations of the consequences of failing to control the spread of LPAI viruses of H5 and H7 subtypes. It therefore seems desirable to control LPAI viruses of H5 and H7 subtype in poultry to reduce the probability of a mutation to HPAI occurring. This in turn may require redefining statutory AI. There appears to be three options:
1. Retain the current definition with locally imposed restrictions to limit the spread of LPAI of H5 and H7 subtypes.
2. Define statutory AI as an infection of birds/poultry with any AI virus of H5 or H7 subtype.
3. Define statutory AI as any infection with AI virus of H5 or H7 subtype, but modify the control measures imposed for different categories of virus and/or different types of host.

Both the EU Scientific Committee on Animal Health and Animal Welfare in 2000 (Scientific Committee on Animal Health and Animal Welfare 2000) and the OIE *ad hoc* Committee on AI in 2002 (OIE 2002) recommended that relevant legislative processes concerned with control or trade should be extended to all infections of poultry with either H5 or H7 viruses.

Keywords: Avian influenza; definition; cleavage site; European Union; OIE

[#] Virology Department, Veterinary Laboratories Agency Weybridge, Addlestone, Surrey KT15 3NB, United Kingdom. E-mail: d.j.alexander@vla.defra.gsi.gov.uk

R. S. Schrijver and G. Koch (eds.), Avian Influenza, 103–112.

Introduction

The first attempt at a universally acceptable definition of what should constitute avian influenza (AI) for which statutory control measures and trading restrictions should apply was agreed at the First International Symposium on Avian Influenza held in Beltsville, USA in 1981 (Bankowski 1982). Until that time definitions used in different countries for 'fowl plague' and 'fowl-plague virus' were extremely variable. It had been known since 1959 that highly virulent AI viruses for poultry could be of two different haemagglutinin subtypes (H7 and H5) and from the early 1970s that not all viruses of these subtypes were necessarily virulent for poultry (Beard and Easterday 1973). Nevertheless, many countries had historical definitions essentially based on identification of viruses as of H7 subtype or the presence of H7 antibodies. The 1981 definition was considered a rational step forward and with subsequent modifications, taking into account the greater understanding of the molecular basis of pathogenicity, it evolved into the current OIE (Office International des Epizooties) definition quoted below.

At the First International Symposium it was recommended that the term 'fowl plague' should be replaced by 'highly pathogenic avian influenza' (HPAI).

Molecular basis of pathogenicity

For all influenza-A viruses the haemagglutinin glycoprotein is produced as a precursor, HA0, which requires post-translational cleavage by host proteases before it is functional and virus particles are infectious (Rott 1992). The HA0 precursor proteins of avian influenza viruses of low virulence for poultry have a single arginine at the cleavage site and another basic amino acid at position -3 or -4. These viruses are limited to cleavage by extracellular host proteases such as trypsin-like enzymes and thus restricted to replication at sites in the host where such enzymes are found, i.e. the respiratory and intestinal tracts. HPAI viruses possess multiple basic amino acids (arginine and lysine) at their HA0 cleavage sites either as a result of apparent insertion or apparent substitution (Senne et al. 1996; Vey et al. 1992; Wood et al. 1993) and appear to be cleavable by (a) ubiquitous protease(s), probably one or more proprotein-processing subtilisin-related endoproteases of which furin is the leading candidate (Stieneke-Grober et al. 1992). HPAI viruses are able to replicate throughout a susceptible avian host, damaging vital organs and tissues, which results in disease and usually rapid death (Rott 1992).

Emergence of highly pathogenic avian influenza

Avian influenza viruses, including those of H5 or H7 subtype, isolated from free-living birds are invariably of low virulence for poultry. Apart from the deaths of large numbers of terns in South Africa in 1961 (Becker 1966), from which A/tern/South Africa/61 (H5N3) was isolated, isolations of HPAI viruses from free-living birds have been associated with contact with infected poultry, usually as a result of surveillance of birds trapped or found dead on or near infected premises. In addition, results of phylogenetic studies of H7 subtype viruses indicate that HPAI viruses do not constitute a separate phylogenetic lineage or lineages, but appear to arise from non-pathogenic strains (Banks et al. 2000; Rohm et al. 1995). Similarly phylogenetic analyses of the preceding LPAI H7N1 isolates and the subsequent HPAI H7N1 isolates in Italy in 1999-2000 indicated evolution from one to the other (Banks et al.

2001). These empirical findings are supported by the in-vitro selection of mutants virulent for chickens from a LPAI H7 virus (Li, Orlich and Rott 1990).

Theories of the molecular basis for the mutation of avian influenza subtype H5 and H7 viruses from low to high virulence in poultry have been put forward by Garcia et al. (1996) and Perdue et al. (1997). Essentially it is proposed that spontaneous duplication of purine triplets results in the insertion of basic amino acids at the HA0 cleavage site and that this occurs due to a transcription fault. The assumption is that this transcription fault occurs more readily in chickens or turkeys than in free-living bird hosts. As pointed out by Perdue et al. (1997) this may not be the only mechanism by which HPAI viruses arise, as some appear to result from nucleotide substitution rather than insertion while others (including the 1999-2000 Italian H7N1 HPAI virus) have insertions without repeating nucleotides. In addition, the H7N3 HPAI virus responsible for the outbreak in Chile in 2002 appears to be somewhat unique. The extremely virulent virus was reported to have a 10-amino-acid insert at the cleavage site giving the unusual motif PEKPKTCSPLSRCRETR*GLF, which does not seem wholly compatible with the need for multiple basic amino acids. The virus is also unique in that the insert appears to have arisen by intergenic recombination between the HA gene and the nucleocapsid gene of the progenitor LPAI virus that had also been isolated (Suarez et al. 2003).

Current definitions

Office International des Epizooties

The following definition for viruses that cause HPAI is taken from the Manual of Standards for Diagnostic Tests and Vaccines 2000 (Alexander 2000)

"*a) Any influenza virus that is lethal for six, seven or eight of eight 4- to 8-week-old susceptible chickens within 10 days following intravenous inoculation with 0.2 ml of a 1/10 dilution of a bacteria-free, infective allantoic fluid.*

b) The following additional test is required if the isolate kills from one to five chickens but is not of the H5 or H7 subtype: growth of the virus in cell culture with cytopathogenic effect or plaque formation in the absence of trypsin. If no growth is observed, the isolate is considered not to be a HPAI isolate.

c) For all H5 and H7 viruses of low pathogenicity and for other viruses, if growth is observed in cell culture without trypsin, the amino acid sequence of the connecting peptide of the haemagglutinin must be determined. If the sequence is similar to that observed for other HPAI isolates, the isolate being tested will be considered to be highly pathogenic."

European Union

European Union (EU) legislation on avian influenza is contained in Council Directive 92/40/EEC (CEC 1992). The disease is defined as follows in Annex III of the directive;

"For the purpose of the diagnostic procedures for the confirmation and differential diagnosis of avian influenza the following definition shall apply.

'Avian influenza' means an infection of poultry caused by any influenza A virus which has an intravenous pathogenicity index[1] in six-week-old chickens greater than 1.2 or any infection with influenza A viruses of H5 or H7 subtype for which nucleotide sequencing has demonstrated the presence of multiple basic amino acids at the cleavage site of the haemagglutinin."

The differences between the two definitions are slight in terms of assessing virus virulence. The decision by the EU to use the Intravenous Pathogenicity Index (IVPI) test means that disease as well as death is assessed, but this involves some subjectivity in reading the test. In practice viruses have qualified by either definitions or neither.

These definitions, formulated over ten years ago, were aimed at including viruses that were overtly virulent in in-vivo tests *and* those that had the potential to become virulent. At that time the only virus known to have mutated to virulence was the one responsible for the 1983/84 Pennsylvania panzootic. In this epizootic viruses isolated at the beginning of AI infections of poultry in Pennsylvania were of low virulence for chickens although possessing multiple basic amino acids at the cleavage site (Kawaoka, Naeve and Webster 1984; Kawaoka et al. 1987). These early viruses possessed a carbohydrate chain close to the cleavage site in the three-dimensional structure of the HA molecule that was absent in the later HPAI isolates. The inference is that the presence of this carbohydrate chain prevented access of the ubiquitous host protease(s), but not trypsin-like enzymes, to the cleavage site and when lost the potential virulence of the virus was realized. This mechanism has not been seen in other viruses. However, the inclusion in these internationally accepted definitions of *potentially* virulent viruses does set a precedent for future definitions.

It should also be noted that both definitions allow the confirmation of HPAI by sequencing the amino acids at the HA0 cleavage site, but in-vivo tests are still required to confirm a virus is LPAI. This is important as RT-PCR primers may well identify only the consensus population or show a better 'fit' with LPAI virus and not detect the presence of HPAI virus in a mixed population of LPAI and HPAI viruses.

Reasons for reviewing the definition

The current theories and the accumulating evidence suggest that HPAI viruses arise from H5 or H7 LPAI viruses infecting chickens and turkeys and that when viruses of these subtypes spread from free-living birds there is always a potential that they may become virulent. However, at present we are unable to predict when and if this will occur. Presumably in outbreaks of HPAI such as that occurring in England in 1991 (Alexander et al. 1993), in which only a single house of turkeys was affected,

---

[1] The intravenous pathogenicity index (IVPI) is the mean score per bird per daily observation over 10 days of 10 six-week-old chickens inoculated intravenously with the virus under test when birds are scored: Score 0 = normal, Score 1 = sick, Score 2 = very sick or paralysed, Score 3 = dead. An IVPI = 0 means that no signs were seen in the 10-day observation period. An IVPI = 3 means that all birds died within 24 hours.

the mutation happens very quickly after introduction. In Australia in 1976 there was evidence of limited spread before mutation took place (Westbury 1997). Whereas in Pennsylvania in 1983 (Webster and Kawaoka 1988), Mexico in 1993/94 (Campos-Lopez, Rivera-Cruz and Irastorza-Enrich 1996; Villarreal and Flores 1997) and Italy 1999/2000 (Capua et al. 2000) there had been extensive outbreaks of LPAI for a considerable period of time before the emergence of HPAI. If it is assumed that mutation to virulence is a random event, then it seems logical that the longer the presence and greater the spread in poultry the more likely it is that HPAI virus will emerge. It would therefore seem a reasonable policy to reduce the spread and presence of LPAI viruses of H5 and H7 subtype in poultry to limit the probability of a mutational event occurring.

Policies pursued, either locally or nationally for different LPAI outbreaks in recent years have varied enormously; ranging from none, through reliance on biosecurity with or without vaccination, voluntary depopulation/slaughter, to a full stamping-out policy, or combinations of these strategies.

In 1998 outbreaks of LPAI caused by virus of H7N7 subtype occurred on the island of Ireland in the Republic of Ireland (29 outbreaks) and in Northern Ireland (3 outbreaks). In both countries the potential to mutate to HPAI viruses and the potential public-health risks were considered serious threats by regulatory authorities and industry. The spread of virus was successfully eliminated by a programme of biosecurity measures, voluntary slaughter, early marketing, cleaning and disinfection and extensive surveillance (Campbell and De Geus 1999; Graham, McCullough and Connor 1999). Similarly, outbreaks of H5 or H7 LPAI in the US have often been controlled by strict biosecurity measures and voluntary depopulation (Eckroade 1997; Senne, Swayne and Suarez 2003). In Utah in 1995 strict biosecurity measures were combined with vaccination (Halvorson et al. 1997). Straightforward stamping out was applied to H5 or H7 LPAI outbreaks in Belgium in 1999 (H5N2), Germany 2001 and Virginia 2002 (H7N2) (review Capua and Alexander 2004) and more recently in Denmark 2003 (H5N7 in ducks). In Italy a 'DIVA' (differentiating infected from vaccinated animals) vaccination strategy was employed with the re-emergence of H7N1 LPAI in 2000 and the H7N3 outbreaks in 2003 (Capua and Alexander 2004), but it should be emphasized that this strategy involves stamping out vaccinated flocks shown to have been infected with the field LPAI virus. All these strategies appear to have been successful in controlling the spread of LPAI.

In contrast, in Italy in 1999 LPAI H7 virus continued to spread despite the recommendation of strict biosecurity regimens, with the emergence of HPAI virus in December 1999 after 199 confirmed LPAI outbreaks (Capua and Marangon 2000).

Many factors appear to influence the ability to control LPAI solely by the application of biosecurity measures including: the degree of spread prior to notification, the population density of poultry farms, the degree of integration and the economic pressures on poultry farmers. The situations in Italy in 1999 and Mexico in 1993/4 are lessons that failure to control LPAI virus spread *will* result in the emergence of HPAI and further complicate the control of the more pathogenic disease. Attempts to control LPAI infections with H5 or H7 viruses without any statutory instrument in place or the ability to pay compensation for birds slaughtered voluntarily may not prove successful.

Chapter 12

Controlling H5 and H7 virus infections

If it is accepted that greater statutory control of H5 and H7 LPAI viruses is necessary to avoid probable emergence of HPAI viruses then the options are relatively limited. The apparent choices are:

1. Retain the current definition with a recommendation that countries impose restrictions to limit the spread of LPAI of H5 and H7 subtypes.

This option essentially maintains the status quo, in that in recent years most countries/states have reacted to try and limit infections of LPAI H5 and H7 viruses when they have occurred in poultry. It has proved successful in some countries and unsuccessful in others.

2. Define statutory AI as an infection of birds/poultry with any AI virus of H5 or H7 subtype.

This option follows the precedent in present definitions of slaughter of birds infected with potentially HPAI viruses (see above), since it is currently thought that all H5 or H7 LPAI viruses may mutate to virulence. The added advantages of this option are that diagnosis of both LPAI and HPAI is greatly simplified and would result in quicker implementation than the current definition as it requires neither in-vivo testing or sequencing of the amino acids at the HA cleavage site.

There are however several disadvantages. There is currently lack of knowledge of the prevalence of H5 and H7 virus infections of poultry, especially species other than turkeys and chickens. In the EU during 2003 member states have been carrying out point prevalence surveillance studies in poultry in an attempt to address this lack of knowledge. There may well be reluctance among farmers to consider slaughter of birds showing few, if any, signs and this could lead to failure to investigate mild respiratory disease or even to covering up infections with LPAI. Some decision would have to be made on whether to treat species such as commercial ducks differently to turkeys and chickens. There is no evidence that H5 and H7 LPAI viruses are likely to mutate while infecting ducks and the prevalence of LPAI viruses of these subtypes could be high in commercial ducks in some countries (Shortridge 1999).

3. Define statutory AI as any infection with AI virus of H5 or H7 subtype, but modify the control measures imposed for different categories of virus and/or different types of host.

This option is intermediate to options 1 and 2. It is envisaged that there would be a legal requirement for the notification of all H5 and H7 infections to the regulatory authorities and there would be statutory imposition of control measures. However, although the presence of HPAI virus would require stamping out, lesser measure could be imposed for LPAI virus infections. Such measures would need to be carefully considered and specified, but could include: voluntary slaughter or early marketing, stringent defined biosecurity measures, epizootiological tracing and surveillance. Possibly infections of commercial ducks could be controlled differently, but the need to prevent spread to other poultry would be paramount.

Conclusions

The EU Scientific Committee on Animal Health and Animal Welfare was asked by the EU Commission to reconsider the definition of AI requiring statutory control and recommended that the current control measures laid down in Council Directive 92/40/EEC should be extended to all infections with either H5 or H7 viruses (Scientific Committee on Animal Health and Animal Welfare 2000). A very similar definition was put forward in an OIE draft chapter for the OIE International Animal Health Code (OIE 2002).

To date there has been considerable debate on the desirability of making this change for control or trade reasons, which is continuing. It was not in the terms of reference of this paper to review the emerging public-health implications of AI infections of poultry, but this may well have a future impact, and any decision based on scientific or poultry-industry criteria may be completely nullified by public-health concerns and public opinion.

References

Alexander, D.J., 2000. Highly pathogenic avian influenza. *In: OIE manual of standards for diagnostic tests and vaccines*. 4th edn. World Organisation for Animal Health OIE, Paris, 212-220.

Alexander, D.J., Lister, S.A., Johnson, M.J., et al., 1993. An outbreak of highly pathogenic avian influenza in turkeys in Great Britain in 1991. *Veterinary Record,* 132 (21), 535-536.

Bankowski, R.A. (ed.) 1982. *Proceedings of the 1st international symposium on avian influenza, 1981*. Carter Comp., Richmond.

Banks, J., Speidel, E.C., McCauley, J.W., et al., 2000. Phylogenetic analysis of H7 haemagglutinin subtype influenza A viruses. *Archives of Virology,* 145 (5), 1047-1058.

Banks, J., Speidel, E.S., Moore, E., et al., 2001. Changes in the haemagglutinin and the neuraminidase genes prior to the emergence of highly pathogenic H7N1 avian influenza viruses in Italy. *Archives of Virology,* 146 (5), 963-973.

Beard, C.W. and Easterday, B.C., 1973. A-Turkey-Oregon-71, an avirulent influenza isolate with the hemagglutinin of fowl plague virus. *Avian Diseases,* 17 (1), 173-181.

Becker, W.B., 1966. The isolation and classification of Tern virus: influenza A-Tern South Africa-1961. *Journal of Hygiene,* 64 (3), 309-320.

Campbell, G. and De Geus, H., 1999. Non-pathogenic avian influenza in Ireland in 1998. *In: Proceedings of the joint fifth annual meetings of the National Newcastle Disease and Avian Influenza Laboratories of countries of the European Union, Vienna 1998.* 13-15.

Campos-Lopez, H., Rivera-Cruz, E. and Irastorza-Enrich, M., 1996. Situacion y perspectivas del programa de erradicacon de la influenza aviar en Mexico. *In: Proceedings of the 45th western poultry disease conference, May 1996, Cancun, Mexico.* 13-16.

Capua, I. and Alexander, D.J., 2004. Avian influenza: recent developments. *Avian Pathology,* 33 (4), 393-404.

Capua, I. and Marangon, S., 2000. The avian influenza epidemic in Italy, 1999-2000: a review. *Avian Pathology,* 29 (4), 289-294.

Capua, I., Mutinelli, F., Marangon, S., et al., 2000. H7N1 avian influenza in Italy (1999 to 2000) in intensively reared chickens and turkeys. *Avian Pathology*, 29 (6), 537-543.

CEC, 1992. Council Directive 92/40/EEC of 19 May 1992 introducing Community measures for the control of avian influenza. *Official Journal of the European Commission* (L 167, 22/06/1992), 1-16.

Eckroade, R.J., 1997. Comment. *In:* Slemons, R.D. ed. *Proceedings of the 4th international symposium on avian influenza, held May 29-31, 1997*. US Animal Health Association, Georgia Center for Continuing Education, The University of Georgia, Athens, 55.

Garcia, M., Crawford, J.M., Latimer, J.W., et al., 1996. Heterogeneity in the haemagglutinin gene and emergence of the highly pathogenic phenotype among recent H5N2 avian influenza viruses from Mexico. *Journal of General Virology*, 77 (part 7), 1493-1504.

Graham, D., McCullough, S. and Connor, T., 1999. Avian influenzas in Northern Ireland: current situation. *In: Proceedings of the joint fifth annual meetings of the National Newcastle Disease and Avian Influenza Laboratories of Countries of the European Union, Vienna 1998*. 18-19.

Halvorson, D.A., Frame, D.D., Friendshuh, A.J., et al., 1997. Outbreaks of low pathogenicity avian influenza in USA. *In:* Slemons, R.D. ed. *Proceedings of the 4th international symposium on avian influenza, held May 29-31, 1997*. US Animal Health Association, Georgia Center for Continuing Education, The University of Georgia, Athens, 36-46.

Kawaoka, Y., Naeve, C.W. and Webster, R.G., 1984. Is virulence of H5N2 influenza viruses in chickens associated with loss of carbohydrate from the hemagglutinin? *Virology*, 139 (2), 303-316.

Kawaoka, Y., Nestorowicz, A., Alexander, D.J., et al., 1987. Molecular analyses of the hemagglutinin genes of H5 influenza viruses: origin of a virulent turkey strain. *Virology*, 158 (1), 218-227.

Li, S.Q., Orlich, M. and Rott, R., 1990. Generation of seal influenza virus variants pathogenic for chickens, because of hemagglutinin cleavage site changes. *Journal of Virology*, 64 (7), 3297-3303.

OIE, 2002. Report of the ad hoc group on avian influenza. *In: Preliminary final report of the meeting of the OIE International Animal Health Code Commission, Rio de Janeiro (Brazil), 25 November-5 December 2002*. Office International des Epizooties, Paris, 151-177.

Perdue, M., Crawford, J., Garcia, M., et al., 1997. Occurrence and possible mechanisms of cleavage site insertions in the avian influenza hemagglutinin gene. *In:* Slemons, R.D. ed. *Proceedings of the 4th international symposium on avian influenza, held May 29-31, 1997*. US Animal Health Association, Georgia Center for Continuing Education, The University of Georgia, Athens, 182-193.

Rohm, C., Horimoto, T., Kawaoka, Y., et al., 1995. Do hemagglutinin genes of highly pathogenic avian influenza viruses constitute unique phylogenetic lineages? *Virology*, 209 (2), 664-670.

Rott, R., 1992. The pathogenic determinant of influenza virus. *Veterinary Microbiology*, 33 (1/4), 303-310.

Scientific Committee on Animal Health and Animal Welfare, 2000. *The definition of avian influenza and The use of vaccination against avian influenza.* European Commission, Scientific Committee on Animal Health and Animal Welfare. [http://europa.eu.int/comm/food/fs/sc/scah/out45_en.pdf]

Senne, D.A., Panigrahy, B., Kawaoka, Y., et al., 1996. Survey of the hemagglutinin (HA) cleavage site sequence of H5 and H7 avian influenza viruses: amino acid sequence at the HA cleavage site as a marker of pathogenicity potential. *Avian Diseases,* 40 (2), 425-437.

Senne, D.A., Swayne, D.E. and Suarez, D.L., 2003. Avian influenza in the Western Hemisphere including the Pacific Islands and Australia. *Avian Diseases,* 47 (Special issue), 798-805.

Shortridge, K.F., 1999. Poultry and the influenza H5N1 outbreak in Hong Kong, 1997: abridged chronology and virus isolation. *Vaccine,* 17 (suppl. 1), S26-S29.

Stieneke-Grober, A., Vey, M., Angliker, H., et al., 1992. Influenza virus hemagglutinin with multibasic cleavage site is activated by furin, a subtilisin-like endoprotease. *Embo Journal,* 11 (7), 2407-2414.

Suarez, D.L., Senne, D.A., Banks, J., et al., 2003. A shift in virulence from low pathogenic to highly pathogenic avian influenza for the H7N3 virus responsible for outbreaks of disease in poultry in Chile appears to be the result of intergenic recombination between the haemagglutinin and NP genes. *In: Abstracts of the XII international conference on negative strand RNA viruses, Pisa, Italy, June 14-19th, 2003.* 164.

Vey, M., Orlich, M., Adler, S., et al., 1992. Hemagglutinin activation of pathogenic avian influenza viruses of serotype H7 requires the protease recognition motif R-X-K/R-R. *Virology,* 188 (1), 408-413.

Villarreal, C.L. and Flores, A.O., 1997. The Mexican avian influenza H5N2 outbreak. *In:* Slemons, R.D. ed. *Proceedings of the 4th international symposium on avian influenza, held May 29-31, 1997.* US Animal Health Association, Georgia Center for Continuing Education, The University of Georgia, Athens, 18-22.

Webster, R.G. and Kawaoka, Y., 1988. Avian influenza. *Critical Reviews in Poultry Biology,* 1, 211-246.

Westbury, H.A., 1997. History of high pathogenic avian influenza in Australia and the H7N3 outbreak 1995. *In:* Slemons, R.D. ed. *Proceedings of the 4th international symposium on avian influenza, held May 29-31, 1997.* US Animal Health Association, Georgia Center for Continuing Education, The University of Georgia, Athens, 23-30.

Wood, G.W., McCauley, J.W., Bashiruddin, J.B., et al., 1993. Deduced amino acid sequences at the haemagglutinin cleavage site of avian influenza A viruses of H5 and H7 subtypes. *Archives of Virology,* 130 (1/2), 209-217.

13

Avian influenza control strategies in the United States of America

D.E. Swayne# and B.L. Akey##

Abstract

Prevention, control and eradication are three different goals or outcomes for dealing with avian influenza (AI) outbreaks in commercial poultry of the USA. These goals are achieved through various strategies developed using components of biosecurity (prevention or reduction in exposure), surveillance and diagnostics, elimination of infected poultry, decreasing host susceptibility to the virus (vaccination or host genetics) and education. However, the success of any developed strategy has depended on industry–government trust, co-operation and interaction. The preferred outcome for HPAI has been stamping out, for which the federal government has regulatory authority to declare an emergency and do immediate eradication of HPAI, and pay indemnities. For H5and H7LPAI, strategies vary from an immediate control plan followed by an intermediate to long-term strategy of eradication. The state governments have regulatory authority over H5 and H7 LPAI, but work co-operatively with USDA in joint programmes. Stamping out has been occasionally used as has controlled marketing, but inconsistently, indemnities have been funded by the state governments and the poultry industries, and less frequently by USDA. Vaccines have been occasionally used but require USDA license of the vaccine and approval from both state and federal government before use in the field. Non-H5 and -H7 LPAI generally follow a preventive programme, such as H1N1 swine-influenza vaccination for turkey breeders. In other situations, control and eradication strategies are followed but regulatory authority is lacking for USDA. Most programmes for LPAI are voluntary and industry-driven.

Keywords: avian influenza; biosecurity; control; diagnosis; euthanasia; vaccination

Introduction

Poultry and poultry products are a major source of high-quality protein as human food and the per-capita consumption has been increasing around the world during the past two decades. In 2002, the broiler-meat production in the world was 52 million metric tons (MT) with United States (12 million MT), China (9.5 million MT) and European Union (6.4 million MT) being the top three producers (FASonline 2003a). Exports of broiler meat accounted for a little over 10% of total production (5.7 million metric tons), of which the United States (2.2 million MT – 39%), Brazil (1.6 million

---

Southeast Poultry Research Laboratory, Agricultural Research Service, United States Department of Agriculture, 934 College Station Rd., Athens, Georgia 30605, USA. E-mail: dswayne@seprl.usda.gov
Division of Animal Industry, New York Department of Agriculture and Markets, 1 Winners Circle, Albany, New York 12235-0001 USA

R. S. Schrijver and G. Koch (eds.), Avian Influenza, 113–130.
© 2005 *Springer. Printed in the Netherlands.*

MT – 28%) and European Union (0.9 million MT – 16%) were the leading exporters with 83% of the market. Similarly, world turkey-meat production was 4.9 million MT with the United States (2.5 million MT), European Union (1.7 million MT) and Brazil (0.2 million MT) being the primary producers (FASonline 2003b). World exports of turkey meat are 636,000 MT with European Union (285,000 MT), United States (199,000 MT) and Brazil (85,000 MT) being the principle exporters.

Maintaining poultry free from high-pathogenicity (HP) avian influenza (AI), a list-A disease as defined by the Office International des Epizooties (OIE), is essential to continue trade in poultry and poultry products between nations (Alexander 1997). In addition, some countries specify freedom from avian influenza viruses of low pathogenicity (LP), principally H5 and H7, before importing poultry and poultry products. Over the last decade, the impact of trade on national animal-health policies has increased. As a result, national policies have focused not only on disease control as a national need, but also on the expectation for continuing or expanding exports. Therefore, national control strategies are impacted by the right of importing nations to protect their own poultry populations from introduction of catastrophic diseases, such as HPAI, through implementation of sanitary and health standards to assure freedom in the importing product from such disease-causing agents. However, at times, some countries have used non-tariff trade barriers as strategies to protect domestic poultry production when legitimate sanitary and health issues do not exist. In implementing non-tariff trade barriers, only scientifically sound risk assessments should be used to identify real threats for disease introduction and distinguish these from perceived threats or political protective intents. For example, in 2002, the US had embargoes imposed by a trading partner against pasteurized egg products from the Midwest after the US reported H7N2 LPAI in the state of Virginia. This incident was not a legitimate trade barrier because: 1) H7N2 LPAI virus was not covered under the World Trade Organization by the OIE international health code, 2) the infection was compartmentalized to a small geographic region in the Eastern United States, and 3) the pasteurization process used on the product would inactivate any influenza virus that might have been present.

Strategy components for dealing with avian influenza

In dealing with avian influenza in the United States, different strategies have been developed and used. Each strategy has been designed with one of three different goals or outcomes in mind: preventing the introduction of avian influenza into poultry, controlling losses by minimizing the negative economic impact of avian influenza when present, or total elimination of avian influenza (eradication), especially the highly pathogenic form. These goals are achieved through various strategies developed using universal components that include: 1) biosecurity (management procedures to prevent introduction or escape of AI virus), 2) diagnostics and surveillance (detection of AI virus infections), 3) elimination of AI-virus-infected poultry, 4) decreasing host susceptibility to the virus (vaccination or host genetics), and 5) education. In developing and implementing specific strategies, multiple factors are considered and include virus strain (pathotype and haemagglutinin subtype), which poultry commodity or commodities are affected, the density of poultry in a geographic area, the demands of export markets, federal verses state regulatory authority, and availability of financial compensation. The success of any strategy is dependent on industry–government trust, co-operation and interaction. In the USA, the federal government has regulatory authority over eradication of HPAI viruses, while the state governments have jurisdiction over LPAI viruses.

Biosecurity

Biosecurity is the utilization of best management practices to reduce the risk of introducing avian influenza virus in a poultry house, farm or operation, either for the initial case or secondary cases in an *ongoing* outbreak, or preventing movement of avian influenza virus off a premise containing infected birds to a new site. In most situations, these practices focus on preventing movement of the avian influenza virus on contaminated equipment, clothing and shoes off of farms with infected birds; preventing movement of infected poultry or their by-products (e.g. manure); or preventing exposure of poultry to wild waterfowl. Farm quarantine is included in biosecurity practices. In many instances, practising biosecurity means controlling the movement of people including restrictions to minimize the number of visitors to farms. This is best achieved by restricting *inbound* and *outbound* movements through circumferential fencing of the farm and locking of the gates, or even better, a manned guard shack to ensure adherence to biosecurity policies (Figure 1). Other high-risk activities must be managed by proper cleaning and disinfection (C&D) of equipment

Figure 1. Guard house which adds an additional layer of biosecurity by screening out visitors and assuring proper biosecurity procedures are followed for entry and exit

shared between farms, decontamination of clothing and shoes of workers (preferably having work clothing and shoes left on the farm with laundering locally), having employees dedicated to one farm, and having strict rules prohibiting employees from owning backyard or recreational poultry or from visiting other poultry farms or establishments. For shared employees such as vaccination crews, catch crews, feed truck drivers, service personnel etc., they must diligently practise C&D of equipment (including vehicles), clothing and shoes, and minimize their exposure to the birds. Ideally, poultry farms should be of low density in a geographic area to reduce the ease of farm-to-farm transmission (Capua and Marangon 2003). Regional biosecurity plans to co-ordinate movement of LPAI-recovered birds to slaughter; disposal of AI-contaminated manure and AI-infected carcasses, C&D of farms and repopulation have been developed and used in California and Minnesota with success (Halvorson 1998; Cardona 2003). In the Virginia H7N2 LPAI outbreak (2002), the movement of daily mortality off the farm to a rendering facility was associated with spread of avian influenza from farm to farm (Akey 2003). This necessitates finding alternative methods for disposal of daily mortality such as on-farm composting or implementation of revised biosecurity procedures to prevent rendering trucks from entering the farm. The latter is best achieved by placement of the daily mortality in rigid containers at the end of farm entry roads and practising proper C&D.

The most important aspect of biosecurity is the development of a 'biosecurity culture' on the premise and within the company. Employees will not practise biosecurity unless they understand the procedures and importance of biosecurity, and how they affect the company and ultimately their jobs. An education component is essential, as will be discussed later. In developing a biosecurity culture, the ownership and management must take it seriously and practise biosecurity in order for the employees to abide by biosecurity policies. Similarly, communication of the location of influenza cases between companies is important in developing regional biosecurity plans.

Preventing direct or indirect exposure to AI-virus-infected birds is very important; including preventing exposure to potentially infected wild birds. Between 1978 and 2000, poultry farmers in Minnesota experienced 108 introductions of LPAI viruses of various haemagglutinin and neuraminidase subtypes from migratory ducks into turkeys (Halvorson 2002). Twenty of these introductions were from H5 or H7 LPAI viruses, but none of these mutated to HP as occurred in Chile in 2002 (H7N3), Italy in 1999 (H7N1), Mexico in 1994 (H5N2) and Pennsylvania in 1983 (H5N2), possibly because the industry eliminated the viruses in less than 6 months. These Minnesota cases resulted from close direct contact between seasonal migratory juvenile ducks (September to November) with range-reared turkeys, or usage of AI-virus-contaminated lake or pond water for indoor-reared turkeys. Although the range-reared or semi-confinement method has represented historically less than 5% of turkey production, this minor production method has been the introduction point for LPAI viruses into Minnesota commercial turkeys with disastrous results. The worst outbreak year was 1995 with 178 farms having LPAI-virus-infected turkeys, predominantly the H9N2 subtype. With the H5N1 HPAI poultry outbreak and human infections in Hong Kong, the Minnesota production companies agreed to stop range rearing of turkeys to eliminate introduction of waterfowl LPAI viruses and a potential public-relations problem should an outbreak of LP or HPAI occur in Minnesota. As a result, from 1997-2000 only 28 flocks had infections with LPAI viruses, mostly from swine H1N1 influenza virus. However, with the development and usage of organic standards for poultry, outdoor rearing will increase in popularity and enhance the risk for introduction of avian influenza into farming systems.

Diagnostics and surveillance

The speed with which a new disease is eliminated is dependent upon how rapid the index case is detected and how fast eradication strategies are implemented. The presence of high mortality is suggestive of an exotic disease such as highly pathogenic avian influenza or velogenic Newcastle disease, but a definitive diagnosis requires the identification of the causative agent (Swayne, Senne and Beard 1998; Swayne and Halvorson 2003). Other causes of high mortality include some toxins, water deprivation and heat exhaustion. In addition, the presence of respiratory problems or drops in egg production could be consistent with LPAI, but other more common viral, bacterial, fungal and non-infectious causes need to be excluded. Historically, diagnosis of avian influenza has required isolation in embryonating chicken eggs and identification by antigen testing, but this is a slow laboratory process requiring 1-3 weeks depending on the number of negative back passages performed and the availability of embryonating eggs. In the past 5 years, the United States Department of Agriculture (USDA) has begun using two alternatives to virus isolation and identification for screening and diagnosing AI: 1) direct detection of type-A influenza virus proteins or antigens, and 2) amplification and detection of AI virus genes.

During the Virginia H7N2 LPAI outbreak of 2002, management decisions had to be made rapidly in the field to quarantine flocks based on a reasonable suspicion of AI virus infection, and identify AI-virus-negative flocks to allow low-risk movement to slaughter (Akey 2003). A membrane-bound, sandwich enzyme-linked immunosorbent assay (antigen-ELISA) to detect Type-A influenza virus antigen (Directigen®, Becton Dickinson, Cockeysville, Maryland) was used as a screening test to detect the presence of AI virus in tracheal swabs from poultry. This assay was useful in several situations. First, if poultry flocks were showing clinical signs of respiratory disease or experienced an abrupt drop in egg production, the antigen-detection system was effective as a diagnostic screening tool to identify AI virus infections in poultry. In previous studies, this antigen-ELISA had 100% specificity and 79% sensitivity for detecting AI virus from such poultry sampled during the 1997 H7N2 LPAI outbreak in Pennsylvania (Davison, Ziegler and Eckroade 1998). Second, the test was useful in the Virginia surveillance programme within the quarantine zone to identify infected birds using the daily mortality removed from the farms, and to allow movement of clinically normal flocks to slaughter when the test results were negative (Akey 2003). Similarly, this test is being used in the current H7N2 LPAI outbreak in Connecticut layers by screening chickens from the daily mortality for the presence of AI virus infections. Previous experimental studies have shown the value of the antigen-ELISA to detect AI virus in tracheal and cloacal swabs of chickens showing clinical respiratory disease and in identifying AI virus in allantoic fluid of inoculated embryonating chicken eggs (Slemons and Brugh 1997). However, during the Virginia H7N2 LPAI outbreak, a few farms had false-positive results on the antigen-ELISA when the birds sampled were moderately to severely autolysed. In the 2003 Connecticut H7N2 LPAI outbreak, a similar problem of false-positive results occurred in autolysed birds, possibly resulting from an alkaline-phosphatase reaction produced by saphrophytic bacteria (Mary J. Lis, personal communication). All tracheal samples screened with antigen-ELISA in Virginia and Connecticut H7N2 LPAI outbreak were tested for AI virus by RRT-PCR and virus isolation to confirm the results of the antigen-ELISA.

Although the antigen-ELISA was an effective field-screening test, a rapid more sensitive and specific diagnostic test was need for avian influenza. A USDA co-operative effort between Southeast Poultry Research Laboratory (Athens, Georgia) and National Veterinary Services Laboratories (Ames, Iowa) developed and validated a laboratory-based, one-step, real-time reverse-transcriptase polymerase chain reaction (RRT-PCR) assays for detection of avian influenza virus in field specimens (Spackman et al. 2002). These RRT-PCR tests have been used for diagnosis in three different H7N2 LPAI outbreaks: live poultry markets of New England during 2002, commercial poultry in Virginia during 2002, and commercial layers of Connecticut during 2003.

The RRT-PCR assays utilized a single nucleic-acid extraction system (Rneasy kit, Qiagen, Valencia, California) from transport media containing pooled tracheal or cloacal swabs, hydrolysis primer and probe sets, and a portable thermocycler. The first assay used an influenza virus matrix-gene-specific PCR primer set and a hydrolysis probe designed for a conserved region present in all type-A influenza viruses; whether avian, swine, equine or human. For all samples which were matrix-gene positive, H5- and H7-specific primer sets to conserved regions of the North-American H5 and H7 haemagglutinin gene sequences, respectively, were used in secondary tests. For all assays, the probes used a 6-carboxyfluorescein reporter dye and 6-carboxytetramethylrhodamine quencher dye.

The RRT-PCR assays took less than 3 hours for completion, which includes sample preparation time, and the answer was produced in real time (Figure 2). On a flock or market basis, the RRT-PCR tests performed well compared to virus isolation with sensitivity of 94% for matrix-gene assay and 97% for H7 haemagglutinin-gene assay during the live-poultry market eradication programme (Spackman et al. 2002). The detection limit for the matrix-gene test was 10^{-1} 50% egg infectious doses (EID_{50}) while the H5 and H7 assays detected 10^{1} EID_{50}. These RRT-PCR tests performed equally well in Virginia during the summer of 2002 using tracheal swabs from meat turkeys, turkey breeders, broiler breeders and broilers. In Connecticut during February 2003, the RRT-PCR test was used to detect the H7N2 LPAI virus in Connecticut egg layers. This USDA-developed and -validated RRT-PCR assay is laboratory-based and has been disseminated to several state veterinary diagnostic laboratories within the National Animal Health Laboratory Network. However, hand-held instrumentation is under development that will make the test usable in the field or at pen side (Perdue, Swayne and Suarez 2003).

Figure 2. Portable thermocycler with laptop computer operating system displaying the results of RT-PCR reactions in real-time (Courtesy of David Suarez)

National surveillance for avian influenza in commercial poultry is accomplished through three systems: 1) National Poultry Improvement Plan, a Federal–State–Industry partnership, to certify chicken and turkey breeders flocks as AI free; 2) testing of broiler and meat-turkey flocks for product export to Mexico; and 3) state programmes to detect AI in high-risk areas (Myers et al. 2003). Specifically, these programmes use the agar-gel immunodiffusion (AGID) and two commercial ELISA (IDEXX, Westbrook, Maine; and Synbiotics, San Diego, California) tests that detect antibodies against the nucleoprotein and matrix protein of all Type-A influenza viruses. All positive results from the ELISA tests are confirmed by AGID test. All AGID-positive samples are forwarded to NVSL for haemagglutinin and neuraminidase subtyping. Serological testing has been used to certify an area or farm as free of AI, or during an AI outbreak to determine the extent of the infected zone for quarantine purposes.

Elimination of infected poultry

AI-virus-infected poultry flocks have been eliminated through two systems: 1) controlled marketing of convalescent or recovered flocks, and 2) *on-farm* depopulation and disposal of infected flocks. The method of elimination depends on the production company's procedures; local, state and federal agricultural and environmental regulatory policies; availability of indemnities; and accessibility to different disposal methods.

Historically, some LPAI virus-infected meat-turkey flocks have been allowed to recover from infection and were marketed through routine processing (Halvorson 2002). However, for processing, the convalescent flocks have been handled differently from non-AI-virus-infected flocks; i.e. with processing occurring at the end of the day followed by disinfection of the plant and delivery trucks before the beginning of the next day of transport and processing of AI-virus-negative flocks.

Euthanasia and disposal is the preferred method of eliminating flocks acutely infected with HPAI virus or recovered from such an infection. In addition, in some situations, euthanasia and disposal of LPAI-virus-infected poultry has been used such as in Virginia during the 2002 H7N2 LPAI outbreak. Depopulation requires two processes: 1) rapid, humane euthanasia of large numbers of poultry, and 2) disposal of the carcasses in an environmentally sound way.

Euthanasia

Usage of carbon-dioxide gas is the preferred method for euthanasia, but administration can be a logistic problem. With caged layers, individual birds must be removed from cages and manually placed in large airtight containers and CO_2 added. With small numbers of birds, portable, self-contained euthanasia chambers can be constructed and moved to the site for use (Figure 3) (Webster, Fletcher and Savage 1996). With large numbers of layers, large steel trash containers covered with airtight tarps and filled with CO_2 have been used effectively for euthanasia. For broilers, meat turkeys and breeders, portable panels were used to construct enclosures, the birds were herded into the enclosures and covered with plastic. Upon introduction of CO_2 (0.08 to 0.11 lbs of CO_2/ft^3) (Figure 3), the birds were euthanized in less than 15 minutes (Akey 2003).

Figure 3. Self-contained stainless steel CO_2 chamber for humane euthanasia of poultry

Carcass disposal

Disposal of euthanized birds has utilized various methodologies depending on local circumstances, including local, state and federal environmental laws. During the 2002 H7N2 LPAI outbreak in Virginia, 4.7 million birds on 197 farms were affected and these farms were depopulated (Akey 2003). Elimination of the infected birds utilized various methods including *on-farm* burial, incineration, composting, landfill disposal and controlled marketing (Table 1). **First,** *on-farm* burial was used in the Virginia

Table 1. Elimination methods used in H7N2 LPAI outbreak in Virginia during 2002 (Bruce Akey).

Disposal method	Number of birds	Total (%)
Composting	43,000	0.9
Incineration	641,000	13.4
Landfill	3,103,000	65.5
Controlled marketing	943,000	19.9
On-farm burial	15,000	0.3
Total	4,732,000	100.0

H7N2 LPAI outbreak only on the first affected turkey farm using an emergency permit and a lined-pit system. Future use of *off-farm* burial will be greatly limited because of public concern and environmental regulations designed to prevent contamination of the ground water. **Second,** forced air-curtain, wood-burning incinerators were used in the Virginia H7N2 LPAI outbreak, but on a limited basis because of complaints about smoke, and because of the high cost for fuel and maintenance of equipment. This incineration process cost $500 per ton plus cost for transportation of the carcasses and disposal of the ash. In other situations for poultry disposal, gas-fired incinerators with afterburners have been used because they are designed to meet air-quality emission standards (Figure 4), but high fossil-fuel costs still made incineration one of the most costly options for disposal of poultry carcasses.

Figure 4. Stationary natural-fired incinerator with afterburners used for disposal of poultry carcasses at a research facility

In The Netherlands 2003 H7N7 HPAI outbreak, incineration was the preferred method of disposal with shipping of euthanized birds in sealed trucks to large stationary incineration plants that meet air-quality emission standards. **Third**, a small number of birds in the Virginia H7N2 LPAI outbreak were composted in the houses by a windrow method or in the commercial Ag-Bag® system (Ag-Bag International, Warrenton, Oregon). The latter method cost $13 per ton for the bagging materials, but required the fixed overhead cost of purchasing a specific loader for the system ($50,000). The windrow method of composting has a similar low cost, but requires the addition of an auxiliary carbon source (such as hay or wood-chip litter), proper construction of the compost windrow, control of vermin to prevent removal of infected carcasses, and periodic turning of the compost pile (Murphy 1992). One advantage of the Ag-Bag® system is the lack of any requirement for turning the compost pile because aerobic digestion is maintained by forced-air and not passive ventilation. In experimental studies, H5N2 HPAI virus was totally inactivated at the

121

end of the first 7-10 days of the composting process using a windrow system (Mixson 1992). On some poultry farms, continuous composting of daily mortality is used as a primary disposal method (Merka et al. 1994). On-farm composting of AI-virus-infected carcasses and manure during AI outbreaks has a good potential for more use in future AI or Newcastle-disease eradication efforts. **Fourth**, most birds in the Virginia H7N2 LPAI outbreak were disposed of by discarding in an approved landfill, but the primary location was a 3-hour drive outside the quarantine zone causing the transportation costs to be enormous. The landfill tipping fees varied from $45 to $140 per ton ($730,000 total cost) plus transportation charges and cleaning and disinfecting the transport trucks. **Fifth**, some birds in the Virginia H7N2 LPAI outbreak were marketed. These birds originated from flocks detected at slaughter but which had shown no symptoms before, or from flocks detected early during the outbreak which had recovered from AI infection. They were marketed because at that time landfill space was unavailable for disposal.

The Minnesota AI Control plan is based on recovery of grower cost by marketing the recovered, asymptomatic birds (Halvorson 1998; 2002). This has worked effectively since 1978, and when compared with the cost to control the H7N2 LPAI outbreak in Virginia by stamping out, the 25 years of LPAI in Minnesota (1097 farms) cost growers $22 million versus $130 million to the industry in losses for 2002 Virginia LPAI outbreak (197 farms) plus eradication costs of $82 million. For LPAI, marketing of recovered birds should be given more serious consideration in the future as a method of elimination. In experimental studies using the H7N2 LPAI virus, meat from intranasally inoculated broilers did not contain any AI virus based on virus isolation attempts, and feeding the meat obtained from inoculated chickens did not transmit the LPAI virus to other chickens (David Swayne, unpublished data). By contrast, broilers inoculated intratracheally with H5N2 HPAI virus from 1983 resulted in AI-virus isolation from meat, and feeding of the meat to other broilers resulted in seroconversion to the AI virus. For LPAI, marketing of recovered birds should be given more serious consideration in the future as a method of elimination since the raw meat produce has a negligible potential for transmitting AI viruses. An alternative would be to use meat from infected flocks for further processing and pre-cooked products. This would be effective for both LP and HPAI viruses since they are thermally labile (Swayne and Halvorson 2003).

Rendering of carcasses was an option used in the Italian H7N3 HPAI outbreak (Capua and Mutinelli 2001). However, in the USA, the rendering industry has been reluctant to accept birds from an AI outbreak because of stigma associated with the product source. Although the rendering process will kill the virus, extra precautions must be taken to prevent reintroduction by cross-contamination of transport vehicles.

Decreasing host susceptibility

Vaccination

Vaccination with homologous haemagglutinin AI vaccines has been shown to decrease susceptibility of poultry to infection by avian influenza viruses. Studies with a fowlpox recombinant virus containing an H5 AI-virus gene insert prevented transmission of AI virus between *in-contact* vaccinated chickens (Swayne, Beck and Mickle 1997). In other studies, vaccination with inactivated whole AI-virus or recombinant live AI-virus vaccines prevented clinical disease and mortality, and decreased replication and shedding of the field virus from respiratory and digestive tracts (Swayne 2003). However, vaccines do not completely prevent infection, especially in the field, thus biosecurity practices are essential to prevent spread between vaccinated flocks that may become infected. Reviews have been written covering AI vaccines and should be referred to for more detail (Capua and Marangon 2003; Swayne 2003).

AI vaccines have been developed, licensed and used in the USA during the last 25 years. AI inactivated vaccines have been produced and licensed under both the autogenous and conditional (limited) licensing authorities (Myers and Morgan 1997; Myers et al. 2003). Full licensure has been granted to a fowl poxvirus recombinant containing an H5 AI-gene insert and an inactivated H5 whole-AI-virus vaccine. However, USDA-licensed vaccines can only be used under permit and in an official government control programme. Specifically, usage of H5 and H7 vaccines requires approval by USDA and the state government, but usage of USDA-licensed AI vaccines of the other 13 haemagglutinin subtypes (non-H5 and non-H7) only requires approval of the state government (Myers et al. 2003). AI vaccines have not been routinely used in the USA for AI prevention, control or eradication. Most vaccines have been used in turkeys, but usually only during individual outbreaks and primarily in turkey breeders (Halvorson 2002). However, in the case of turkey breeders, H1N1 influenza vaccine has been used in some states to prevent egg-production drops from infection by H1N1 swine influenza viruses. During 2001, 2,697,000 doses of H1N1 vaccine were used in 5 states (Illinois, Michigan, North Carolina, Minnesota and Ohio) and an additional 100,000 doses of H1N2 autogenous H1N2 vaccine were used in Missouri turkey breeders (Swayne 2001). In the same year, 677,000 doses of H6N2 inactivated AI vaccine were used in California on one layer farm. Since 1978, a variety of different haemagglutinin-subtype AI inactivated vaccines have been used in Minnesota turkey breeders and meat turkeys (Halvorson 1997).

Host resistance
In vitro studies using transfection technologies have shown that Mx genes from some chicken breeds conferred resistance to influenza infection in mouse 3T3 cells, but not from a White leghorn line (Ko et al. 2002). In an in-vivo study, mild differences in virulence of an LPAI virus were noted between specific-pathogen-free White leghorn, commercial White Leghorn and broiler chickens (Swayne et al. 1994). The impact of host genetic selection on resistance to AI virus infections has not been fully determined.

Education
Education is an essential component of any AI prevention, control or eradication strategy. This involves providing information to the industry concerning the biology of avian influenza viruses, how the virus is introduced and spread on farms, and application of methods and practices in biosecurity to prevent introduction of AI virus onto a farm (exclusion biosecurity practices) and, on infected premises, biosecurity practices to prevent the AI virus from leaving (inclusion biosecurity practices). The education process involves all employees in the company who are provided information on what avian influenza is, how it is transmitted, identification and elimination of behaviours that put the farm at risk for AI introduction (e.g. owning backyard poultry, working or visiting on multiple poultry farms), biosecurity procedures to protect the poultry (e.g. farm-dedicated clothing and shoes left on the farm at the end of the workday, employee showering facilities before entry on the farm, C&D of equipment used between farms) and consequences to the company and their jobs if an AI outbreak involves their workplace. Risk communication is essential

between companies when AI-infected premise is identified. In the Minnesota turkey industry this has been accomplished by a telephone and mail alert system to the poultry industry for suspicious and confirmed cases of AI (Poss 1997). The industry then develops and implements a responsible response for eradication of AI (Poss 1997; Halvorson 2002).

Specific strategies used in USA for avian-influenza prevention, control and eradication

HPAI

There have been three outbreaks of HPAI in the USA during the 20th century: 1924-25 fowl plague in live-poultry markets of Northeast cities, and supplying farms in Northeast and upper Midwest; 1929 fowl plague in a few New Jersey farms; and 1983-84 H5N2 HPAI, principally in Pennsylvania, but also limited involvement in Maryland and Virginia (Swayne and Halvorson 2003). The basic goal has been rapid eradication accomplished through quarantine of infected flocks, depopulation of flocks and disposal of carcasses, C&D of equipment and farms, diagnostics and surveillance testing (1983-84) and indemnities paid for destruction of poultry (1983-84). Federal laws and regulations give USDA the authority to declare an animal-health emergence, to quarantine and destroy flocks, and to pay indemnities (Myers et al. 2003). These eradication programmes can be conducted in co-operation with state governments and are outlined in the USDA Emergency Response Plan (APHIS 1998), also called the 'Red Book'. This document provides for the potential future use of vaccines as part of eradication strategies for HPAI.

H5 and H7 LPAI

The goal or outcome for H5 and H7 LPAI has been control or eradication. Strategies to accomplish the goal of eradication or control have varied with the individual situation. Such strategies utilized components of biosecurity, diagnostics and surveillance, elimination of infected birds and in some situations, vaccination. Prior to 1995, use of USDA-licensed AI vaccines only required a decision by the poultry industry and state governments, but in 1995, USDA implemented the requirement of federal approval for field use of USDA-licensed H5 and H7 vaccines (Myers et al. 2003). In most situations, indemnities have not been paid for elimination of H5 and H7 LPAI-infected poultry. Currently, federal regulations do not provide for indemnities to cover H5 and H7 LPAI. In the next few paragraphs, control or eradication strategies for specific outbreaks of H5 and H7 LPAI will be covered, with primary focus on the use or non-use of vaccines, payment or non-payment of indemnities, and elimination methods used. Biosecurity enhancements, quarantine and surveillance are common components used in all the strategies to control or eradicate AI.

Minnesota, 1978-2002

Since 1978, the Minnesota turkey industry has experienced 108 outbreaks with AI viruses in turkeys involving 1097 farms, twenty of these outbreaks were from waterfowl-origin H5 or H7 LPAI viruses (Halvorson 2002). In each instance, the industry implemented an AI-eradication strategy utilizing components of education (includes risk communication within the poultry industry), monitoring, responsible response (includes controlled marketing and enhanced biosecurity practices), and vaccination with the outcome of eradication of AI in less than six months (Halvorson

2002). Prior to 1995, H5 and H7 vaccines were allowed without USDA approval, but since 1995 none have been used. The Minnesota control programme was a voluntary co-operative industry–state programme and did not include any indemnity from the federal or state governments, but relied on marketing of recovered turkeys as financial compensation for participation. Moving or marketing of turkeys during the acute phase of infection, i.e. period of high AI-virus excretion, was reported as a high-risk activity associated with spreading of the virus to other farms; therefore, infected flocks are required to sit on the farm under quarantine for 1-2 weeks before sending to processing.

Utah, 1995

On 26 April 1995, an H7N3 LPAI virus was isolated from commercial meat turkeys in a single production company in the Sanpete Valley, Utah (Halvorson et al. 1997). Control measures implemented included informing growers of the outbreak, enhanced biosecurity, controlled marketing of recovered flocks, and C&D of housing. An H7N3 autogenous inactivated AI vaccine was produced and, beginning on 20 June 1995, uninfected 3-8-week-old turkeys were given a single dose of the vaccine. Within 6 weeks, 150 flocks were vaccinated. Sentinel birds placed at the time of immunization did not seroconvert over the next 6 months. The company concluded the vaccine was effective in reducing susceptibility of turkeys to AI virus and, along with biosecurity measures, ending the outbreak. No state or federal indemnities were paid as compensation.

Live-poultry markets, 1994-2002

The H7N2 AI outbreak in live-poultry markets of the Northeastern USA (1994-2002) and in commercial poultry inVirginia (2002) are covered by Dennis Senne in this volume and will only be discussed below in brief. Between 1994 and 2001, several attempts at eliminating H7N2 LPAI by identifying markets with infected birds, closure of infected markets, eliminating infected birds, and C&D were ineffective (Mullaney 2003). In April 2002, a federal–multiple state co-operative programme was launched with goal of a simultaneous closure of 123 retail markets in six Northeastern states. The owners sold down their bird inventories the day before closure and killed all remaining birds on the day of closure. Each market owner was compensated $3000 for the three days of closure if the establishment handled only poultry, and a $1000 supplement was paid if they handled both poultry and red-meat livestock. The establishments were cleaned and disinfected by the owners and inspected by the task force before being allowed to repopulate with certified AI-free birds.

Pennsylvania, 1996-97 and 2001-2002

In 1996-97, H7N2 LPAI spilled over from the live-poultry markets into 18 commercial layer farms, two commercial layer pullet farms and one commercial meat-turkey farm in Pennsylvania (Davison et al. 2003). The control strategy was placement of quarantine, immediate depopulation with *on-farm* burial or delayed depopulation with landfill burial, C&D of premises, and surveillance. Depopulation was voluntary without any federal compensation on 21 farms. However, on two large layer farms, chickens were allowed to recover and continue to produce market eggs. Within 6 months, AI virus was again recovered from chickens within both flocks. Ultimately, the flocks were depopulated and buried in a landfill. Request by the industry to vaccinate against H7 AI was not approved by USDA because of broiler industry's concerns over a potential trade embargo on broiler-meat exports if vaccine was used. Partial indemnity was paid by the state government and from an industry indemnity pool. In 2001-02, H7N2 LPAI virus spilled over from live-poultry markets to infect broiler breeders on two farms and broilers on five farms in Pennsylvania (Dunn et al. 2003). Several of the broiler farms did partial load-outs and sold birds to wholesalers for live-poultry market distribution. The broiler breeders were euthanized in the houses with CO_2 gas and transported to landfill for disposal. For the broiler flocks, one was marketed, one was euthanized on site with landfill disposal, and three flocks were euthanized and composted within the houses. No vaccine was used and federal indemnity was not paid. Other components of the control programme were quarantine and surveillance. The area was declared AI free within 6 months.

Virginia, 2002

H7N2 LPAI virus moved from live-poultry markets to infect 197 commercial farms of turkey and broiler breeders, meat turkeys and broilers during the spring 2002 in the Shenandoah Valley of Virginia. An eradication programme was undertaken with a USDA–Virginia-State co-operative programme funded by USDA (Akey 2003). Stamping out was the strategy used for eliminating infected birds and, for the depopulation, the Federal government paid indemnity. This unusual action by the Federal government to fund eradication of H7N2 LPAI resulted from the progressive change in the LPAI virus; i.e. between 1994-2002 substitutions of non-basic with basic amino acids near the haemagglutinin proteolytic cleavage site. One additional change by a single nucleotide could alter the genetic code and substitute a fifth basic amino acid at the hemagglutinin proteolytic cleavage site. This genetic change would mean an accompanying change in virulence from LP to HP. The turkey industry requested use of inactivated H7N2 AI vaccine in repopulated turkey breeders, which was approved by USDA, but concerns by the broiler industry of an embargo of meat sales by trading partners resulted in non-approval of vaccination by the state government. Initially, diagnosis was made by virus isolation, but during the outbreak sufficient samples were examined to validate the sensitivity and specificity of the RRT-PCR test for detecting AI virus. During the second half of the outbreak, the RRT-PCR assay became the primary test for diagnosing AI virus and virus isolation became a secondary test. The total federal cost of eradication was $81 million and losses to poultry farmers were in excess of $130 million.

Connecticut, 2003

In February 2003, H7N2 LPAI was diagnosed in chickens within a large layer company in Connecticut (Swayne et al. in press). Over the next 3 months, three farm sites in the company involving 2.9 million layers became infected. USDA requested the owner depopulate the infected flocks, but indemnities were not available. The state government and company developed an alternative control strategy to prevent the infection from spreading and eliminate infected birds over their production cycles with the 1-year goal of eradication. The basic strategy is to isolate the farms through biosecurity practices, increase the immunity in infected layers to a uniform level by a single inactivated H7N2 AI-virus vaccination, over the normal production cycle replace infected layers with twice-vaccinated pullets (H7N3 AI vaccine), and establish a monitoring programme using unvaccinated sentinels and normal daily mortalities for virus detection. A 'DIVA' serological monitoring strategy using neuraminidase is under consideration. As of 12 October 2003, the last H7N2 LPAI virus detected by RRT-PCR was on 26 June 2003. The control strategy will continue through one laying cycle with periodic evaluation of progress. Isolation or detection of AI virus by RRT-PCR will result in re-evaluation of the control strategy and possibly total depopulation.

Rhode Island, 2003

On 26 April 2003, a 32,000-bird mixed layer operation in Rhode Island became positive for AI virus on RRT-PCR test. Seven days earlier, a live-poultry market dealer visited the farm to purchase birds. The farm is under quarantine with no addition of new birds. The farm can continue to sell sanitized eggs, but at the end of lay the farm will be depopulated, C&D and repopulated with AI-free layers.

Concerning control and eradication of H5 and H7 LPAI, specific issues continue to create problems in the development of consistent and effective eradication strategies: 1) lack of federal authority over H5 and H7 LPAI; 2) inconsistent availability of funds for the eradication efforts, especially for indemnities; 3) continued concerns over potential trade embargoes should vaccines be used in a control or eradication programme; 4) the penalty for using vaccines under OIE health code which require twice the length from last positive case to be declared AI-free (6 versus 12 months); and 5) concerns that if H5 and H7 LPAI are made reportable to OIE, this will necessitate immediate eradication strategies using expensive and disruptive stamping-out policies.

Non-H5 and non-H7 LPAI

Prevention, control and eradication have been strategies used for dealing with non-H5 and non-H7 LPAI. In Minnesota since 1978, the same control and eradication strategy has been used with non-H5 and non-H7 LPAI as with H5 and H7 LPAI, except usage of vaccination continues to be a component in non-H5 and non-H7 LPAI eradication strategies. By contrast, with H1N1 swine influenza, prevention has been the preferred strategy in turkey breeders where vaccination is the primary component used in the prevention strategy. In the recent H6N2 LPAI outbreak in California, the Minnesota control and eradication plan was modified and used. However, success was elusive because biosecurity practices on some individual farms were inconsistent. H6N2 inactivated vaccine has been used in layers but not broilers. The outbreak of exotic Newcastle disease in 2002-03 has temporarily delayed the H6N2 LPAI control programme.

Conclusions

The development and implementation of new control programmes for 'reportable AI' will require courageous steps by all countries that are members of OIE. The addition of H5 and H7 LPAI along with HPAI to the list of 'reportable AI' should reduce the number of HPAI outbreaks in the future by providing governments with incentives to control LPAI before it mutates to HP. However, if trading partners use the addition of H5 and H7 LPAI as a non-tariff trade barrier, this will have the opposite of the intended effect by encouraging nations not to report but to hide LPAI, and this could possible lead to increased HPAI outbreaks in the future. Specific steps need to be taken to make H5 and H7 LPAI reportable:

1. OIE needs to embrace the idea that control methods besides stamping out can be effective and less costly for eradicating H5 and H7 LPAI. LP and

HPAI have different pathogenesis of infection, different virus-shed rates and different rates of transmission, thus their risks are different.

2. Acceptance of the compartmentalization concept is critical for developing new control and eradication methods for HPAI and for including H5 and H7 LPAI as 'reportable AI'.

3. Federal–state co-operative control and eradication programmes need to be developed with financial incentives for rapid detection and elimination of index cases of H5 and H7 LPAI.

4. USDA needs legal authority to control H5 and H7 LPAI including financial resources to pay indemnities.

Acknowledgments

The author acknowledges and thanks Mary J. Lis for contributions of information. Issues on AI prevention, control and eradication have been discussed and documented at the 1st-5th International Symposia on Avian Influenza (1981-2002). The proceedings of these 5 symposia are available as a set (paper copy of 5th and CD-ROM of 1st-4th) for $50 from American Association of Avian Pathologists (953 College Station Rd, Athens, Georgia 30602-4875, USA; tel.: +1 706-542-5645, fax: +1 706-542-0249; AAAP@uga.edu).

References

Akey, B.L., 2003. Low-pathogenicity H7N2 avian influenza outbreak in Virginia during 2002. *Avian Diseases,* 47 (special issue), 1099-1103.

Alexander, D.J., 1997. Highly pathogenic avian influenza (fowl plague). *In: OIE Manual of standards for diagnostic tests and vaccines: list A and B diseases of mammals, birds and bees.* 3rd edn. Office International des Epizooties, Paris, 155-160.

APHIS, 1998. *Avian influenza emergency disease guidelines.* Animal Plant Health Inspection Service, USDA, Hyattsville.

Capua, I. and Marangon, S., 2003. The use of vaccination as an option for the control of avian influenza. *Avian Pathology,* 32 (4), 335-343.

Capua, I. and Mutinelli, F., 2001. *A color atlas and text on avian influenza.* Papi Editore, Bologna.

Cardona, C., 2003. The control of avian influenza and exotic Newcastle disease in California. *In: Proceedings of the 54th north central avian disease conference, Cleveland, 21-23 September 2003.* Ohio Agricultural Research and Development Center, Wooster, 9.

Davison, S., Eckroade, R.J., Ziegler, A.F., et al., 2003. A review of the 1996-98 nonpathogenic H7N2 avian influenza outbreak in Pennsylvania. *Avian Diseases,* 47 (special issue), 823-827.

Davison, S., Ziegler, A.F. and Eckroade, R.J., 1998. Comparison of an antigen-capture enzyme immunoassay with virus isolation for avian influenza from field samples. *Avian Diseases,* 42 (4), 791-795.

Dunn, P.A., Wallner-Pendleton, E.A., Lu, H., et al., 2003. Summary of the 2001-02 Pennsylvania H7N2 low pathogenicity avian influenza outbreak in meat type chickens. *Avian Diseases,* 47 (special issue), 812-816.

FASonline, 2003a. *Broiler statistics.* USDA, Foreign Agricultural Service. [http://www.fas.usda.gov/dlp/circular/2003/03-03LP/poultry_sum.pdf]

FASonline, 2003b. *Turkey statistics*. USDA, Foreign Agricultural Service. [http://www.fas.usda.gov/dlp/circular/2003/03-03LP/turk_sum.pdf]

Halvorson, D.A., 1997. Strengths and weaknesses of vaccine as a control tool. *In:* Swayne, D.E. and Slemons, R.D. eds. *Proceedings of the 4th international symposium on avian influenza, held May 29-31, 1997.* US Animal Health Association, Georgia Center for Continuing Education, The University of Georgia, Athens, 223-226.

Halvorson, D.A., 1998. Epidemiology and control of avian influenza in Minnesota. *In: Proceedings of the 47th New England Poultry Health Conference, Portsmouth, New Hampshire, 25-26 March 1998.* New England Poultry Association, Portsmouth, 5-11.

Halvorson, D.A., 2002. Twenty-five years of avian influenza in Minnesota. *In: Proceedings of the 53rd north central avian disease conference, Minneapolis, 6-8 October 2002.* NCADC, Minneapolis, 65-69.

Halvorson, D.A., Frame, D.D., Friendshuh, A.J., et al., 1997. Outbreaks of low pathogenicity avian influenza in USA. *In:* Swayne, D.E. and Slemons, R.D. eds. *Proceedings of the 4th international symposium on avian influenza, held May 29-31, 1997.* US Animal Health Association, Georgia Center for Continuing Education, The University of Georgia, Athens, 36-46.

Ko, J.H., Jin, H.K., Asano, A., et al., 2002. Polymorphisms and differential antiviral activity of the chicken Mx gene. *Genome Research,* 12 (4), 595-601.

Merka, B., Lacy, M., Savage, S., et al., 1994. *Composting poultry mortalities.* The University of Georgia College of Agricultural & Environmental Sciences, Cooperative Extension Service. Circular 819-15. [http://www.ces.uga.edu/pubcd/c819-15w.html]

Mixson, M.A., 1992. Stability/liability of avian influenza viruses in the depopulation process. *In:* Easterday, B.C. ed. *Proceedings of the 3rd international symposium on avian influenza, Madison, Wisconsin, 27-29 May 1992.* US Animal Health Association, Richmond, 155-158.

Mullaney, R., 2003. Live-bird market closure activities in the Northeastern United States. *Avian Diseases,* 47 (Special Issue), 1096-1098.

Murphy, D.W., 1992. Disposal of avian influenza infected poultry by composting. *In:* Easterday, B.C. ed. *Proceedings of the 3rd international symposium on avian influenza, Madison, Wisconsin, 27-29 May 1992.* US Animal Health Association, Richmond, 147-153.

Myers, T.J. and Morgan, A.P., 1997. Policy and guidance for licensure of avian influenza vaccines in the United States. *In:* Swayne, D.E. and Slemons, R.D. eds. *Proceedings of the 4th international symposium on avian influenza, Athens, Georgia, May 29-31, 1997.* Georgia Center for Continuing Education, Athens, 373-378.

Myers, T.J., Rhorer, M.D.A., Clifford, J., et al., 2003. USDA options for regulatory changes to enhance the prevention and control of avian influenza. *Avian Diseases,* 47 (special issue), 982-987.

Perdue, M.L., Swayne, D.E. and Suarez, D.L., 2003. Molecular diagnostics in an insecure world. *Avian Diseases,* 47 (special issue), 1063-1068.

Poss, P.E., 1997. Avian influenza in the turkey industry: the Minnesota model. *In:* Swayne, D.E. and Slemons, R.D. eds. *Proceedings of the 4th international symposium on avian influenza, Athens, Georgia, May 29-31, 1997.* Georgia Center for Continuing Education, Athens, 335-340.

Slemons, R.D. and Brugh, M., 1997. Rapid antigen detection as an aid in early diagnosis and control of avian influenza. *In:* Swayne, D.E. and Slemons, R.D. eds. *Proceedings of the 4th international symposium on avian influenza, Athens, Georgia, May 29-31, 1997.* Georgia Center for Continuing Education, Athens, 313-317.

Spackman, E., Senne, D.A., Myers, T.J., et al., 2002. Development of a real-time reverse transcriptase PCR assay for type A influenza virus and the avian H5 and H7 hemagglutinin subtypes. *Journal of Clinical Microbiology,* 40 (9), 3256-3260.

Swayne, D.E., 2001. Avian influenza vaccine use during 2001. *In: Proceedings of the 104th annual meeting of the U.S. Animal Health Association, Hershey, Pennsylvania, 9-14 October 2001.* US Animal Health Association, Richmond, 469-471.

Swayne, D.E., 2003. Vaccines for list A poultry diseases: emphasis on avian influenza. *In:* Brown, F. and Roth, J.A. eds. *Vaccines for OIE list A and emerging animal diseases: international symposium, Ames, Iowa, USA, 16-18 September, 2002: proceedings.* Karger, Basel, 201-212. Developments in Biologicals no. 114.

Swayne, D.E., Beck, J.R. and Mickle, T.R., 1997. Efficacy of recombinant fowl poxvirus vaccine in protecting chickens against a highly pathogenic Mexican-origin H5N2 avian influenza virus. *Avian Diseases,* 41 (4), 910-922.

Swayne, D.E. and Halvorson, D.A., 2003. Influenza. *In:* Saif, Y.M., Barnes, H.J., Glisson, J.R., et al. eds. *Diseases of poultry.* 11th edn. Iowa State University Press, Ames, IA, 135-160.

Swayne, D.E., Radin, M.J., Hoepf, T.M., et al., 1994. Acute renal failure as the cause of death in chickens following intravenous inoculation with avian influenza virus A/chicken/Alabama/7395/75 (H4N8). *Avian Diseases,* 38 (1), 151-157.

Swayne, D.E., Senne, D.A. and Beard, C.W., 1998. Influenza. *In:* Swayne, D.E., Glisson, J.R., Jackwood, M.W., et al. eds. *Isolation and identification of avian pathogens.* 4 edn. American Association of Avian Pathologist, Kennett Square, 150-155.

Swayne, D.E., Smith, B., Myers, T.J., et al., in press. Avian influenza in the USA. *In: Proceedings of the 38th national meeting on poultry health and processing, Ocean City, Maryland, 22-24 October 2003.* Delmarva Poultry Industry, Georgetown.

Webster, A.B., Fletcher, D.L. and Savage, S.I., 1996. Humane on-farm killing of spent hens. *Journal of Applied Poultry Research,* 5 (2), 191-200.

ECONOMICS OF AVIAN INFLUENZA CONTROL

14

Minimizing the vulnerability of poultry production chains for avian influenza

C.W. Beard#

Even a casual observer of the global avian influenza (AI) situation will likely be quick to arrive at the conclusion that should bring concern to the world's poultry industries. That is, avian influenza is not going to 'go away'. Because it has migratory waterfowl, sea birds, shore birds and perhaps some other wild avian species as its natural hosts, avian influenza is here to stay with its 'built-in' dissemination mechanism: the free-flying birds. The presence of the viruses in the guts of the usually asymptomatic free-flying natural hosts and the shedding of the viruses in their faeces facilitates the widespread perpetuation of AI viruses along with the ever-present threat they present to domestic poultry.

Recent AI experiences in widely different locations such as Hong Kong and New York support the conclusion that live-bird markets (LBMs) can effectively serve as the direct or indirect 'link' for AI between the natural hosts and domestic poultry. The LBMs are usually a collection of birds from many sources and may include many species including waterfowl and some especially susceptible species such as quail. The LBMs are perfect 'links' because they also include commercial poultry, which can facilitate the transfer of virus from the LBMs back to the commercial producers via contaminated transport coops, vehicles and personnel.

The backyard flocks can also include a variety of avian species including domestic waterfowl on open ponds exposed to direct contact with migratory waterfowl, game fowl (fighting cocks) and free-ranging guinea fowl, geese and an untold number of other avian species. They can also participate in the poultry adaptation and transfer of AI to commercial poultry flocks through the LBM system.

There is apparently an increased interest by some of the more recent residents in the United States in both buying their poultry meat at an LBM and having backyard flocks which in some areas of the country are likely to include the frequently transported fighting cocks. Although illegal in all but a single state, the interest in rearing game fowl (fighting cocks) is apparently increasing as a hobby/business in some cultural circles in certain geographic areas of the country. There are no indications that either the LBM system of collecting and marketing dressed poultry or the practice of having backyard flocks (even in dense urban areas) is going to wane. In fact, they appear to be on the increase, which adds to the ever present threat of AI gaining access to commercial poultry flocks.

Whereas there are some poultry diseases for which we might be able to discuss their complete eradication with the reasonable likelihood of being successful, I do not believe avian influenza is one of those. For the reasons mentioned above, AI viruses with their global distribution are here to stay.

Vice President - Research & Technology, U.S. Poultry & Egg Association, 1530 Cooledge Road, Tucker, Georgia 30084 USA. E-mail: cbeard@poultryegg.org

R. S. Schrijver and G. Koch (eds.), Avian Influenza, 133–137.
© 2005 *Springer. Printed in the Netherlands.*

It is usually when the AI viruses apparently 'spill over' from their natural hosts and become adapted to domestic poultry species that they become the cause for concern. The highly pathogenic members of the AI viruses can result in very serious economic losses to commercial poultry. The recent evidence that some of the H5 and H7 AI viruses can be directly transmitted from poultry to humans, causing infection, disease and even death adds more urgency to devising reliable schemes to protect domestic poultry from the ubiquitous avian influenza viruses. If AI was ever to become generally perceived to be a dangerous zoonotic disease, it could have a very negative effect on the consumption of poultry meat and eggs.

Once we can agree that the AI viruses are not likely to 'go away' and are likely to persist in the world for the foreseeable future, we can be much more realistic as we try to devise a plan or plans of action to lessen their negative impact. We know that vaccines certainly offer a means to provide variable levels of protection against AI. Once the identification of the particular haemagglutinin of the AI virus of concern is known, vaccines (inactivated, subunit or vectored) can be of great benefit if properly managed. The use of naturally occurring avirulent AI viruses to protect against virulent members of the same HA subtype has generally been discouraged for two reasons: they can participate in the development of new viruses with unpredictable consequences through reassortment between the virulent and the avirulent viruses, and the live 'vaccine' virus may also experience mutations to become highly pathogenic as well.

Even though there are therapeutic drugs licensed for use in humans against type-A influenza, they do not appear, at least for now, to have practical use in domestic poultry. A published laboratory trial using one of the first of these drugs (amantadine HCl) demonstrated that AI-virus populations exhibiting drug resistance could rapidly emerge with the virus replicating and causing mortality in chickens being medicated with the drug. The rapidly acquired drug resistance was confirmed by in-vitro laboratory studies of the viruses recovered from the affected chickens.

There is another very obvious option to the prevention and control of losses from AI, and that is simply to produce poultry in a way to prevent their exposure to the AI virus. It is easy to make such a statement that could appear naïve to some in that the success of such a simplistic approach depends upon so many variables including: the density and make-up of the poultry population in the area, the proximity of poultry premises to each other, the proximity to concentrations of the free-flying natural hosts (flyways or preserves), the extent and multiplicity of types of domestic poultry in the area and, probably most important of all, the level of knowledge of poultry biosecurity practices and the commitment of the poultry-company management as well as the on-farm personnel to the protection of their flocks from exposure to AI.

The possibility of the intentional introduction of disease agents such as highly pathogenic AI and Newcastle viruses into commercial flocks by those wishing to inflict serious economic and food-supply harm has received considerable attention in the last two years. Some of us do not believe that it would be possible to prevent all if any initial attacks but believe it would be feasible to develop and implement a system of biosecurity that would confine the diseases to the flocks where the diseases were initially introduced, not allowing them to spread throughout the poultry-company complex.

The vertical integration of the broiler industry in the United States has resulted in complexes comprised of hatcheries, feed mills, processing plants, breeder flocks and broiler flocks. The flocks of one company may be interspersed in the same geographic area with flocks owned by other broiler companies. Because of the day-to-day movement of company personnel between their flocks, sometimes at great distances, I believe the existing practice of establishing disease quarantine zones simply on distance from infected flocks during control programmes should be re-examined. I do not believe distances have as much relevance as they once did, because in the modern industry a company service person can visit an infected flock and travel 70-80 km to another company flock within the period of one hour. Such a scenario tells me that the validity of setting up 10-km 'zones' around infected flocks may be of limited value in today's commercial industry. Flock-contact sequences by company and other personnel appear to be a more important consideration than distance in that there is some evidence that airborne transmission between flocks occurs only when houses are in very close proximity. It is only possible to lessen the chances of such disease spread by human movement through a sound and closely adhered to system of biosecurity.

Because we at the U.S. Poultry & Egg Association believe that keeping AI out of flocks through 'biosecurity' is a valid, albeit difficult, option for preventing AI losses, we funded the development and production of an interactive compact disc (CD) training and reference programme entitled "Poultry Disease Risk Management: Practical Biosecurity Resources CD".

The individual primarily responsible for the content of the CD was Dr. J. P. Vaillancourt , a poultry veterinarian at North Carolina State University. He solicited and received input from poultry-industry veterinarians throughout the CD development process utilizing the rapid-response capability and convenience of email. He applied the term 'Dream Team' to that group and used them for rapid editorial and content reviews of biosecurity issues and practices all during the CD development process.

The actual production of the interactive CD was accomplished by Gene Lambert of Paradigm Media in California. He remains the technical resource person for those who may have difficulty operating all the many features of the CD. Users must have a computer with a sound card and CD-reading drive capability. It is not difficult to use, even by a computer neophyte. It can also be used by professionals in group-training sessions.

The CD is being offered to the poultry industry at no cost. It can be obtained by placing an order on the U.S. Poultry & Egg website at www.poultryegg.org. We are attempting to keep records on who receives the CD so future revisions/updates can be sent to them. Approximately 500 CDs have already been mailed out to requestors.

We have concluded that for any programme of biosecurity to be effective, the bird caretakers or growers themselves must be absolutely convinced of its need, informed of the technical basics involved and supportive of the effort. All the biosecurity rules and regulations sent down to the farm from 'those in charge' will be for naught without the complete 'buy in' by the grower/caretakers of the flocks. For example, no poultry-company administrator or government regulator will be observing the poultry house when the caretaker has to get out of bed at midnight to check the brooder operation and house ventilation. That caretaker must believe in the need for and the effectiveness of biosecurity or he/she might be inclined to take no biosecurity precautions before entering the poultry house in the dark of the night. Caretaker education and training must be a basic building block of creating a system of biosecurity that can keep AI out of commercial poultry. It is the hope of the U.S. Poultry & Egg Association that the Biosecurity CD will, through the coming years, help accomplish that difficult goal.

Because the implementation of an effective biosecurity programme in the commercial poultry industry will likely require both operational and facility changes at some level of increased cost, upper management of the poultry company must also be well informed of the need and be supportive of the effort. Large integrated poultry companies are often led by talented administrators with backgrounds either in poultry science or business administration. As with the flock caretakers, they may not fully understand the basic principles of infection, disease and contagion. Because disease transmission is generally due to the direct or indirect contact of susceptible flocks with infected and virus-shedding poultry or even more likely due to the movement of contaminated personnel and/or equipment, there is a great need to educate all poultry-industry personnel on diseases and their causative agents. Once they have that basic understanding, individuals will be able to appreciate the need for biosecurity practices and therefore will be more likely to be supportive of the biosecurity effort.

There are many operations that are the responsibility of the poultry-company management and not the farm caretakers where there is a great need for attention to biosecurity. From the placement of chicks, flock supervision, feed delivery to catching/hauling, the integrator management has a long list of operations where the neglect of sound biosecurity can be very costly. Any operations that include farm visits can be particularly risky without biosecurity. The collection of semen and the artificial insemination of turkey hens is a good example as is the placement of new pullets and removal of spent fowl in a very large multi-age table-egg operation. To implement a really effective programme of biosecurity will first require a significant education/training effort at all levels in a company; this can only be successfully accomplished with a prolonged and continuing training programme. Employee turnover and retraining of employees and caretakers will require that biosecurity training become a routine part of company training and indoctrination programmes.

The Biosecurity CD that we have produced will hopefully be proven to be a convenient, interactive and easily updated training aid. During the workshop I have presented a PowerPoint 'look' at the CD and its capabilities prepared for promoting the use of the CD by Dr. Vaillancourt, and I have shown a few minutes of the Biosecurity CD.

I am not trying here to convince you that we in the United States are claiming that we have made the startling new 'discovery' that biosecurity is the way to prevent commercial poultry losses from AI and other infectious poultry diseases. We are aware that the realization has been obvious to all of us for many years. However, with the expanding LBM system, the increased interest by more of our residents in having backyard flocks, and the continuing growth of a diverse and, in some locations, a crowded poultry industry, it was our belief that the level of biosecurity practiced in our industry needed to be drastically improved. It is our hope that this Biosecurity CD, with the advantages of its inherent technological features, will help achieve that goal.

It is our goal that the broad disease-prevention benefits of an expanded and improved level of biosecurity can, if properly implemented and maintained, be so significant over time to offset, partially or completely, any additional costs associated with it. After all, biosecurity is mostly a 'people' thing: their understanding of diseases and their transmission, their dedication to complying with biosecurity practices, and the extent to which regulators can be successful in controlling the inherent risks to commercial poultry that accompany the LBM systems and backyard (game fowl) poultry operations.

There are some parts of the commercial industry that should be put on the 'top of the list' for improved biosecurity. The primary breeders are already convinced of the importance of biosecurity and have made great strides to implement it in their companies. There is an obvious need for biosecurity to be improved at the multiplier or parent flock level within companies. These are very valuable birds and the capital investment alone would be ample justification alone for a company to invest in biosecurity as 'insurance' against disease losses. The very large million- and multimillion-hen table-egg complexes also represent a huge bird investment and are other examples where improved biosecurity could represent a economically defensible form of insurance against disease losses. In fact, there may not be any other reasonable alternatives to protecting such large numbers of caged layers against the ravages of AI.

I believe the concept of biosecurity practices reducing the likelihood of AI losses is a sound approach. The remaining question is whether or not those involved will have the interest and the will to pursue that route. After all, if they don't join in with conviction and enthusiasm, it is not likely to be successful. I appreciate the opportunity to have been a part of this AI workshop and welcome future input on how we might improve the Biosecurity CD.

15

Economics of controlling avian influenza epidemics

M.A.P.M. van Asseldonk[#], M.P.M. Meuwissen[#], M.C.M. Mourits[##] and R.B.M. Huirne[#, ##]

Abstract

Livestock epidemics like avian influenza (AI) may inevitably affect many farms at the same time, causing severe economic losses for governments, farmers and all other participants of the poultry production chain. This paper gives an overview of the various economic consequences incurred by AI epidemics. The funding of these losses is discussed, making a distinction between the compensation currently regulated by the EU or national government and the additional financing possibilities by means of insurance or co-financing systems. A general economic analysis is performed to explore the required level of insurance or co-financing rates for the various EU member states to cover the expected AI losses. The level of rates turned out to vary widely among the individual EU countries.

Keywords: contagious animal diseases; avian influenza; economic consequences; compensation; EU; direct losses; consequential losses; co-financing levy; insurance premium

Introduction

The European Union (EU) aims at assuring a high level of animal health and animal welfare without compromising the functioning of the internal market (Mission Statement DG Health and Consumer Protection, *Homepage DG SANCO*). To achieve this goal the EU follows a strategy of non-vaccination for most highly contagious animal diseases. To control possible outbreaks, the EU member states have adopted regular control measures, consisting of stamping out of infected herds, pre-emptive slaughter of contact herds, and the establishment of protection (> 3 km) and surveillance (> 10 km) zones.

During the last decade, a number of large outbreaks of contagious livestock diseases occurred throughout the EU, including foot and mouth disease (FMD), classical swine fever (CSF) and highly pathogenic avian influenza (AI) (*OIE classification of diseases: animal diseases data (website)*). These had a devastating veterinary impact: FMD and CSF caused the slaughter of millions of cattle, sheep and pigs (*OIE classification of diseases: animal diseases data (website)*). For example, during the last epizootic of AI in The Netherlands, around 30 million animals were slaughtered, i.e., half of the Dutch poultry population (*Homepage LNV*). The economic losses due to FMD in the UK in 2001, to CSF in The Netherlands in 1997

[#] Institute for Risk Management in Agriculture (IRMA), Wageningen University, Hollandseweg 1, 6707 KN Wageningen, The Netherlands
[##] Business Economics, Wageningen University, Hollandseweg 1, 6706 KN Wageningen, The Netherlands

R. S. Schrijver and G. Koch (eds.), Avian Influenza, 139–148.
© 2005 *Springer. Printed in the Netherlands.*

and to AI in Italy in 1999-2000 were estimated at 12,500 million Euro, 2,300 mEuro and 500 mEuro, respectively. These figures demonstrate that the economic consequences of highly contagious animal diseases can be enormous.

This paper gives an overview of the various economic consequences incurred by contagious animal disease in general and AI in particular. Subsequently, the funding of these losses is discussed, making a distinction between the compensation currently regulated by the EU and the additional financing possibilities by means of insurance or co-financing systems.

Estimating the insurance premium or a co-financing levy is very complicated. In the final part of the paper a general economic analysis is introduced to explore the required level of insurance or co-financing rates for the various EU member states to cover the losses of an AI outbreak.

Control measures to contain avian influenza outbreaks in the EU

Within the EU, measures to control and eradicate AI have been harmonized by EU legislation (Directive 92/40/EEC). All suspected cases of AI within EU member states must be investigated, while appropriate measures are taken in case of confirmation of highly pathogenic AI. To limit the spread, infected poultry must be killed in a humane way and disposed of safely. Feeding stuffs, contaminated equipment and manure must be destroyed or treated to inactivate the virus.

To prevent further spread of disease the veterinary authorities are required to put in place movement restrictions immediately on the affected holdings and on all farms in a radius of at least 10km around these holdings, the so-called surveillance zone. If necessary, stamping-out measures can also be extended to poultry farms in the vicinity of, or which have had dangerous contacts with, infected farms (pre-emptive slaughter).

In accordance with EU legislation, all member states have AI contingency plans in place to ensure that the most appropriate measures are immediately implemented. Depending on the severity of the epidemic, national governments can take *additional control measures*, such as the pre-emptive slaughter of all flocks within a certain radius of infected and detected flocks. During the 2003 AI epidemic in The Netherlands, the Dutch government decided, for instance, on a temporary nation-wide standstill for the transport of live poultry. Also, flocks within a 1-km radius of detected or suspected flocks were pre-emptively slaughtered. In a further stage of the epidemic this pre-emptive slaughter measure was even more intensified to create poultry-free buffer zones.

If movement restriction zones lead to severe *animal welfare problems* on the farms (e.g. farms with broilers), so-called welfare slaughter is generally applied. With stamping out and pre-emptive slaughter, buildings are completely emptied (i.e. depopulated). With welfare slaughter, buildings may only become partly empty. All animals in stamping-out, pre-emptive slaughter and welfare slaughter programmes are destroyed and rendered.

In accordance with Directive 92/40/EEC, emergency vaccination against AI may be used to supplement the control measures carried out after confirmation of the disease. Birds vaccinated against the subtype corresponding to the one which is circulating are protected against the worst effects of AI.

Economic losses caused by AI epidemics

Economic losses incurred by epidemics of contagious livestock diseases like AI can be divided into various categories. An intuitive distinction can be made between *direct losses* and indirect or *consequential losses*. Direct losses refer to the costs of the execution of the eradication campaign reflected by, for instance, the value of destroyed animals and the costs of organizational aspects such as the monitoring of farms in restriction zones. Consequential losses are the 'long-term' consequences due to movement restrictions and market disruptions. They can be divided into five categories (see also Meuwissen et al. 1999):

(1) *Business interruption*: Business interruption occurs because farm buildings become (partly) empty due to stamping out, pre-emptive slaughter or welfare slaughter, and stay empty until movement restrictions are lifted. On farms that are empty, losses from business interruption can be limited by an alternative application of idle production factors; farmers can for instance renovate their stables, temporarily seek another job, etc.

(2) *Losses related to established movement restriction zones*: farms in movement restriction zones face (long) periods in which animals (such as broilers) and manure cannot be transported from the farm. These periods are characterized by animal welfare problems, extra feeding costs and emergency measures for storage of manure.

(3) *Repopulation of the farm*. These losses include losses due to extra weeks with empty buildings (for example because new layers are not readily available) and extra costs of animal health problems. These losses thus do *not* refer to the costs of buying a new flock; government compensation for the slaughtered flock is generally sufficient to buy back a flock of equal quality.

(4) *Losses from emergency vaccination:* given a situation in which vaccinated animals are destroyed, losses might arise from the above categories (business interruption, repopulation costs). However, for reasons of social acceptability, the rendering of vaccinated animals is under debate. With future epidemics, meat from vaccinated animals may be destined to the local market, which likely leads to extra costs and/or lower prices.

(5) *Price effects*. Livestock epidemics can have a rather severe impact on prices. The size and duration of the impact depends on aspects such as the size of the epidemic (duration, size of restricted area), reactions of other countries (closure of borders, increased production) and whether vaccination is applied (which generally leads to long periods of export limitations). Price effects depend to a large extent on the fact whether a country in which an outbreak occurs is an importing or exporting country with respect to products (e.g., eggs) involved in the export limitations. For exporting countries these price effects may result in enormous losses, exceeding the direct losses many times.

Note that other parts of the agricultural supply chain (e.g., breeding organizations, feed mills, slaughter houses, processing industry, transportation companies) are also affected economically (sometimes positively, mostly negatively). Furthermore, outbreaks of contagious animal diseases can also have serious effects on the other parts of the economy as a whole because of the side effects of disease control measures (e.g. tourism) and interactions between economic sectors (falling prices for livestock products favour consumers). These effects are not further discussed.

Compensation payments

Direct losses

Governments (national and European) generally bear the largest part of the direct losses. The European Union compensates part of the direct losses incurred. The veterinary budget of the European Union refunds in most cases 50% of the costs of compulsory and pre-emptive slaughter, 70% of the costs of welfare slaughter, and 50% of the organizational costs. In case of an approved emergency vaccination programme, the EU also provides funding for vaccines and vaccination campaigns.

The national compensation strategy of direct costs varies among EU member states. While some members states finance the direct losses from the national budget (e.g., UK and Denmark), most member states have set up some form of statutory system to co-finance the direct losses (e.g., Germany and The Netherlands). These public–private co-financing schemes have a compulsory fund structure in which all farmers pay a levy. Risk financing by means of a levy system is based on pooling over time within the livestock sector. Payments to the fund can be organized through up-front payments (deposit) or through assessment payments after an epidemic, or both. However, most member states have opted for an assessment approach. These latter systems have no annually fixed levies. The government will finance the compensation payments in advance. The input of the government will however be repaid over the following years. Therefore, after an epidemic, the levy is set according to the amount that the government paid in advance for the sector. In an 'up-front' programme levies are deposited in a fund before an outbreak. However, the size of the payments (reservations) is uncertain and thus there is also the possibility that assessment payments will be necessary (Van Asseldonk, Meuwissen and Huirne 2004).

In case of co-financing to complement the public part, the amount that is financed by the sector can be proportional or non-proportional, or both. If risks are shared between the sector and the national government by means of a proportional contract (i.e., pro rata contract) the levy is specified as a fraction of the coverage. With non-proportional contracts, the national government indemnifies only claims in excess of a particular threshold.

In The Netherlands the current co-finance contract is based on a private bank-guarantee system (i.e., an up-front levy system). The Dutch guarantee system is in the form of a non-proportional contract. Within the system the government can withdraw capital without prior approval of the farming sector from a private bank to (co-) finance actual losses. Any capital provided by the bank is paid back with interest by the primary sector through assessment payments over a certain time horizon.

The Dutch co-financing system became operational in 2000 by the establishment of the so-called 'National Animal Health Fund'. Individual farmers contribute to this fund by paying a fixed levy per animal or animal product. Adjustments of the levy could be made yearly, depending on the direct costs incurred by eradication in a particular year. Through this system of levies the Dutch poultry-farming sector (approx. 6400 commercial flocks) guarantees a maximal contribution to this fund of 11.3 million Euro in the five year period of 2000-2005. Similar agreements have been made with the pig and cattle sector, each accounting for a maximal contribution of 225 mEuro over the same time period. The Dutch government only compensates direct losses that exceed these predefined contributions, which was obviously the case with respect to the control costs of the 2003 AI epidemic. Total direct control costs of this epidemic equalled an amount of 270 mEuro (Homepage LNV). Based on the

Dutch compensation system, these costs were covered for approximately 50% by the EU, 4% by the poultry industry and 46% by the Dutch government.

Consequential losses

The consequential (long-term) losses of the Dutch 2003 AI epidemic are expected to exceed three to four times the direct costs as a result of the serious disruption of production capacity (i.e. half of the poultry population was (pre-emptively) slaughtered) and of the Dutch export markets (Tacken et al. 2003). However, with respect to these consequential losses there is no financial compensation of the European or national government. These losses are completely borne by the farmers and other participants of the Dutch production chain involved.

Only a limited number of EU member states apply free public disaster assistance or compensate above the value of the animals that are compulsorily culled to cover part of the consequential losses. Public–private partnerships in the sense that national governments subsidize a consequential-loss policy are scarce. Insurance schemes could, therefore, be considered a relevant compensation option for consequential losses. A standard insurance scheme is financed through fixed advance payments (e.g., premiums). Insurance capacity is provided through various layers of own retention and reinsurance. However, in the case of an assessment mutual, the company has the right to assess their clients for losses and expenses via initial premiums or after a loss occurs (additional premiums) or both.

Few private insurance schemes exist on the European market to cover the risk of consequential losses as a result of epidemics in livestock. Those that do exist are either extensions of general livestock insurance policies or specific policies of stock insurers and mutual insurers. Many standard livestock insurance policies in Europe indemnify farmers for animal losses as a result of illness and accidents, but some have been extended, sometimes as an option, to cover at least part of the consequential losses from animal disease epidemics (Van Asseldonk, Meuwissen and Huirne 2004).

Premium or up-front levy setting (hence called rates) is very complicated and should be based on a profound knowledge of all factors included. With knowledge of the most important risk factors, the probability of an outbreak and its economic consequences, the level of the premium or levy can be estimated (Van Asseldonk, Meuwissen and Huirne 2004; Meuwissen, Van Asseldonk and Huirne 2003).

Exploring the level of AI rates for various EU member states

In this section the basis of a general economic analysis is introduced, by which the level of insurance premiums or co-finance levies for various EU members can be estimated. The economic analysis is based on expectations with respect to the potential spread of AI, by placing the consequences in the specific context of the country in which the outbreak might occur.

Such an analysis typically requires many epidemiological (e.g., disease spread, herd structure and animal intensity) and economic (e.g., income per animal, output prices, import/export position of country) data on the outbreak. In most countries, these basic data are not available. Therefore, a questionnaire was developed to collect these required data on introduction and spread by means of expert elicitation. The questionnaire was sent in 2002 to the Chief Veterinary Officers (CVOs) of the member states of the regional Commission Europe of the OIE. The subjective responses of the CVOs were used to provide a first insight into economic consequences of AI epidemics in the participating countries.

The following countries participated in the study: Austria, Belgium, Cyprus, Czech Republic, Estonia, Finland, France, Germany, Greece, Hungary, Ireland, Israel, Italy, Latvia, Luxembourg, The Netherlands, Norway, Portugal, Turkey, Ukraine and United Kingdom (Wilpshaar et al. 2002).

The questionnaire: Eliciting expert opinions on AI epidemics per country

Historical data about the chance of introduction of AI are very limited in most countries. Since outbreaks of AI occur irregularly in time and place it is difficult to derive general properties and predictive values. Also the probability distribution describing the possible spread is difficult to ascertain. Because of this lack of data availability, the analysis used in this research is based on elicited subjective expert knowledge. Tree-point estimates that completely specify the so-called triangular probability distribution (asking for minimum, most likely and maximum values) were elicited to derive information concerning the chance of an outbreak, the number of infected farms, duration and radius of restriction zones. Solely on the basis of these numbers an estimation of the outbreak can be calculated (Van Asseldonk, Meuwissen and Huirne 2004).

As explained before, the above data for this research were collected by a questionnaire that was sent to all CVOs of the European member states of the OIE. The response rate was about 44% (i.e. 21 questionnaires). Thirty-six per cent of the returned questionnaires (i.e., 7 questionnaires) could not be used for the analysis because they were only partly completed and therefore not useful for calculations. So, consequently, the questionnaires of 14 countries were used in the final analysis.

Monte Carlo simulation

A Monte Carlo simulation model is constructed in order to obtain insight into the annual-loss distribution (Hardaker, Huirne and Anderson 1997). Monte Carlo simulation is considered an appropriate and very flexible method to investigate aspects that are stochastic of nature, such as livestock epidemics. Including the possibility of these types of events in a simulation model is an important technique in risk analysis. Risks are thereby incorporated as probability distributions. In this study, the problem situation is analysed by creating a stochastic simulation model, which is manipulated by input modification with respect to the different scenarios or decisions. The applied sampling technique is Monte Carlo sampling in which random numbers are sampled from *a priori* specified distributions, i.e. stochastic simulation. At each iteration, randomly drawn numbers from specified distributions are used representing a possible combination of values that could occur. Combining the results of each iteration will lead to a distribution of output values, reflecting a realistic aspect of chance.

In the Monte Carlo simulation model, a Poisson distribution reflects the uncertainty about the introduction of an epidemic in a specific year. Epidemic and ultimately economic consequences are reflected by triangular distributions, with parameters referring to the most likely, minimum and maximum values derived from the experts through the questionnaire. Results are based on 5000 iterations.

For calculating the economic consequences the following additional assumptions had to be made:

(1) For each infected farm, three farms are slaughtered pre-emptively.
(2) All affected farms (i.e. all farms that are infected, pre-emptively slaughtered and/or located in a restriction zone) face restrictions for the whole duration of the

epidemic (i.e., there are no temporarily removals of restrictions for part of the farms).

More details of the epidemiological and economic models used for our analyses are published by Horst et al. (1999) and Meuwissen et al. (1997; 2003).

Economic analysis based on questionnaire results

The CVOs were asked about the expected occurrence and size of AI epidemics in their country for the next five years (i.e., 2003 - 2008). Table 1 shows the averages of the most likely, minimum and maximum estimated values of the occurrence of an outbreak (using all 14 questionnaires). It also shows the two extreme estimations, referred to as 'optimistic' and 'pessimistic'. As demonstrated by Table 1, AI is expected most likely to occur, on average, 0.86 times per country in the next 5 years. Of the 14 surveyed countries, the most optimistic individual-country estimation corresponded with the most likely, minimum and maximum number of outbreaks in 5 years of 0, 0 and 1, respectively. The country with the most pessimistic estimation reflected a most likely, minimum and maximum number of outbreaks of (respectively) 2, 0 and 5 (Table 1) in 5 years.

Table 1. Expected number of AI outbreaks per country for the years 2003 - 2008

	Most likely (all countries; n=14)	Optimistic (n=1)	Pessimistic (n=1)
AI most likely	0.86	0	2
AI minimum	0.23	0	0
AI maximum	3.08	1	5

In Table 2 the most likely, minimum and maximum estimations for the size of AI epidemics are shown. Included were the number of farms infected, the duration of an epidemic (expressed in days) and the radius of restriction zones (in km). The latter refers to the total area that is expected to be confronted with restrictive measures.

Table 2. Average expected size of AI epidemics for the period 2003-2008 and the most optimistic and most pessimistic individual scenarios (most likely values, and the minimum and maximum values between brackets)

	Most likely (n=14)	Optimistic (n=1)	Pessimistic (n=1)
Number of poultry farms infected	4 (1-32)	1 (1-2)	15 (5-200)
Duration of epidemic (days)	33 (22-95)	5 (3-10)	90 (60-340)
Radius of affected area (km)	18 (8-50)	3 (3-10)	50 (10-200)

The average most likely value of the number of farms that will be affected is 4 with a minimum of 1 and a maximum of 32 (during the 5-year period). The duration of the epidemic is estimated to last 33 days with a minimum of 22 days and a maximum of 95 days. The radius of the affected area is 18 km with a minimum of 8 km and a maximum of 50 km. Again the most optimistic and most pessimistic estimated individual-country scenarios are shown (Table 2).

Direct and consequential losses are obtained by combining the epidemiological estimations of each country with the country-specific financial parameters. Table 3 shows the annual total-loss distributions (summation of direct losses and consequential losses) in million Euros, resulting from the simulation model for AI (5000 iterations). Data per epidemic are aggregated into annual data at the country level by considering the number of epidemics in a certain year and the losses per epidemic. (Countries are reported anonymously.)

Table 3. Total losses per country per year resulting from AI epidemics expressed in million euros and in per cent (%) of the average animal value (average, 0.75 and 0.95 percentile (pc) values)

Country	Million Euros			Per cent of animal value		
	Average	0.75 pc	0.95 pc	Average	0.75 pc	0.95 pc
A	25.4	0.0	179.2	0.94	0.0	6.61
B	11.6	0.0	2.4	19.46	0.0	3.94
C	0.0*	0.0	0.2	0.30	0.0	2.37
D	0.1	0.0	0.0	0.24	0.0	0.0
E	28.8	0.0	206.5	2.23	0.0	15.97
F	0.0*	0.0	0.2	0.15	0.0	1.03
G	0.1	0.2	0.4	0.09	0.16	0.38
H	1.5	2.8	6.5	0.44	0.82	1.92
I	0.0*	0.0	0.0*	0.02	0.0	0.08
J	0.0*	0.0	0.0	0.08	0.0	0.0
K	0.0*	0.0	0.0*	0.11	0.0	0.09
L	0.0*	0.0	0.0	0.05	0.0	0.0
M	0.5	0.0	0.0	0.48	0.0	0.0
N	47.0	0.0	345.0	8.40	0.0	61.71

* Values are not exactly zero but decimals are not sufficient to represent

The results of Table 3 show that the average expected total losses for AI vary from almost € 0.0 million (country C, F, I, J, K and L) to € 47.0 million per year (country N). The 0.75 and 0.95 percentile values indicate that the distributions of the annual total costs are skewed. The 0.75 percentile values demonstrate that – independent of the country surveyed – in 75 per cent of the cases, the estimated annual losses of an AI outbreak are less than € 3mln. However, for the countries A, E and N, there is a 5% chance that these annual losses exceed € 100 mln (see 0.95 percentile values).

Given the estimated loss distribution, including the part that is compensated by the EU, the level of insurance/co-finance rates can be evaluated. The estimated rates are expressed in per cent of the animal value. The spread in rates originates form the number and severity of epidemics occurring. The second part of Table 3 shows the annual distribution of these estimated rates. Again data are aggregated into annual data at country level by considering the number of epidemics in a certain year and the losses per epidemic.

The average expected total losses for AI per year vary from 0.02% of the (country-specific) average poultry value (country I) to 19.46% (country B). The percentile values of the annual rates demonstrate, once more, a rather skewed distribution.

Discussion and conclusion

Poultry production regions can be divided on the basis of the degree of risk to experience an AI epidemic. A risk classification can be based on several criteria such as border regulations (e.g. sanitary controls, hygienic measures, number of live animal imports/exports), flock or animal density, and natural borders (rivers, mountains, seas). These criteria fluctuate widely across the surveyed countries, resulting in varying estimations with respect to the possible chance of an AI outbreak (Table 1) and subsequent epidemiological consequences (Table 2). Along with the export/import situation of a country, these variations result in substantial differences between the country-specific expected rates to cover the economic losses of an AI outbreak (Table 3).

These differences will complicate the formulation of a possible EU-wide insurance to cover the non-compensated part of the losses. A major problem in insuring losses from epidemic diseases is that the epidemic-disease risk is systemic, meaning that many of those who are insured can suffer losses at the same time. Private-sector insurance companies find it difficult to maintain adequate financial reserves and to obtain sufficient reinsurance capacity (i.e. insurance for the insurance company) when dealing with risks that have systemic characteristics. However, recent developments in capital markets provide opportunities to enlarge the reinsurance capacity (Miranda and Glauber 1997). Partnerships between the public and private sectors may furthermore attract some financial involvement from government.

Some form of public–private partnership is also important when considering prevention measures to reduce the risks of an outbreak. A partnership is, for instance, important to prevent delaying debates about prevention and control strategies; some (additional) control measures may seem cost-reducing at the time they are taken, but they may lead to significantly higher total losses (for instance due to enlargement of export bans). Also, a partnership for risk financing is relevant to reduce moral hazard of governments; many epidemics can either be prevented or magnified by government policies (or lack thereof). Having governments financially responsible for some losses might be incentive for them to put into place appropriate animal-disease control measures (Meuwissen, Van Asseldonk and Huirne 2003).

The described economic analysis should be considered a pilot study; the quality of the country-specific results depends on the quality and completeness of the underlying data. For some member states there are not enough data of sufficient quality to perform such an analysis. Lack of detailed data on the location of animals hampers epidemiological calculations that are needed to further define the effects of regional concentration of animal densities. User-friendly software, linked to sound scientific methods, is now available but their application is – at present – severely limited due to the lack of detailed information that is needed as input to these systems (Wilpshaar et al. 2002).

References

Hardaker, J.B., Huirne, R.B.M. and Anderson, J.R., 1997. *Coping with risk in agriculture.* CAB International, Wallingford.

Homepage DG SANCO. European Commission Health and Consumer Protection Directorate-General (DG SANCO) (website). [http://www.europa.eu.int/comm/dgs/health_consumer/index_en.htm]

Homepage LNV. Ministerie van Landbouw, Natuur en Voedselkwaliteit (website). [http://www.minlnv.nl]

Horst, H.S., Dijkhuizen, A.A., Huirne, R.B.M., et al., 1999. Monte Carlo simulation of virus introduction into the Netherlands. *Preventive Veterinary Medicine,* 41 (2/3), 209-229.

Meuwissen, M.P., Horst, S.H., Huirne, R.B., et al., 1999. A model to estimate the financial consequences of classical swine fever outbreaks: principles and outcomes. *Preventive Veterinary Medicine,* 42 (3/4), 249-270.

Meuwissen, M.P.M., Horst, H.S., Huirne, R.B.M., et al., 1997. *Schade verzekerd!? : een haalbaarheidsstudie naar risico-kwantificering en verzekering van veewetziekten.* Wageningen, Landbouwuniversiteit.

Meuwissen, M.P.M., Van Asseldonk, M.A.P.M. and Huirne, R.B.M., 2003. Alternative risk financing instruments for swine epidemics. *Agricultural Systems,* 75 (2/3), 305-322.

Miranda, M.J. and Glauber, J.W., 1997. Systemic risk, reinsurance, and the failure of crop insurance markets. *American Journal of Agricultural Economics,* 79 (1), 206-215.

OIE classification of diseases: animal diseases data (website). Office International des Epizootie OIE. [http://www.oie.int/eng/maladies/en_classification.htm]

Tacken, G.M.L., Van Leeuwen, M.G.A., Koole, B., et al., 2003. *Ketenconsequenties van de uitbraak van vogelpest.* LEI, Den Haag. Rapport LEI no. 6.03.06.

Van Asseldonk, M.A.P.M., Meuwissen, M.P.M. and Huirne, R.B.M., 2004. Risk-financing strategies to manage epidemic animal disease costs in the UK. *Farm Management,* 11 (12), 686-697.

Wilpshaar, H., Meuwissen, M.P.M., Tomassen, F.H.M., et al., 2002. *Economic decision making to prevent the spread of infectious animal diseases.* Proceedings of 20th Conference of the O.I.E. Regional Commission for Europe, September 2002, Finland.

Concluding remarks

This book should speak for itself. However, we felt it necessary to mention the issues that came out the discussion as issues that should be addressed with the highest priority:

The virus
- elucidate the biological relevance of the genetic variation

The monitoring
- improve diagnostic tests to facilitate cost-efficient mass monitoring of LPAI in poultry

The spread
- study the role of wildlife birds in generating and spreading of new variants

The control
- exchange experience in control measures such as mass killing of birds and mass vaccination campaigns, to prevent unnecessary animal or human suffering

The organizations
- organize structural co-operation between FAO, WHO and OIE on avian influenza

The regulations
- reduce the time needed before regulations are adapted to the newest scientific insights

The tools
-explore antivirals
-improve marker vaccines.

We consider the meeting successful thanks to the following conditions:

- the subject was focused
- there were experts from different fields
- there were experts from various countries
- it was a small group
- participants joined in an open debate in a pleasant atmosphere
- the conference was very productive.

We trust it has brought the solution to the avian influenza problem just a little bit nearer!

List of participants

Alexander, Dennis	Virology Department, Veterinary Laboratories Agency Weybridge, Addlestone, UK
Beard, Charles	US Poultry and Egg Association, Tucker, Georgia, USA
Capua, Ilaria	OIE and National Reference Laboratory for Newcastle Disease and Avian Influenza, Istituto Zooprofilattico Sperimentale delle Venezie, Legnaro, Padova, Italy
Cargill, Peter	Merial Animal Health, Hereford, UK
De Jong, Mart	Animal Sciences Group, Wageningen UR, Lelystad, The Netherlands
De Vries, Tjep	Animal Health Service, Deventer, The Netherlands
De Wit, Jac	Animal Health Service, Deventer, The Netherlands
Ellis, Trevor	Tai Lung Veterinary Laboratory, Agriculture, Fisheries and Conservation Department, Kowloon, Shueng Shui, New Territories, Hong Kong, China
Enting, Ina	Research Institute for Animal Husbandry, Animal Sciences Group, Wageningen UR, Lelystad, The Netherlands
Fouchier, Ron	Department of Virology, Erasmus Medical Centre, Rotterdam, The Netherlands
Ge, Lan	Animal Sciences Group, Wageningen UR, Wageningen, The Netherlands
Hermans, Tia	Alterra, Wageningen UR, Wageningen, The Netherlands
Huirne, Ruud	Animal Sciences Group, Wageningen UR, Wageningen, The Netherlands
Jorgensen, Pol	Danish Veterinary Institute, Århus, Denmark
Kaleta, Erhard	Klinik für Vögel, Reptilien, Amphibien und Fische, Giessen, Germany
Koch, Guus	Central Institute for Disease Control, Wageningen UR, Lelystad, The Netherlands
Koopmans, Marion	RIVM, Bilthoven, The Netherlands
Laddomada, Alberto	European Commission, Brussels, Belgium
Le gall-recule, Ghislaine	Agence Française de Sécurité Sanitaire des Aliments, Ploufragan, France
Leenstra, Ferry	Animal Sciences Group, Wageningen UR, Lelystad, The Netherlands
Marangon, Stefano	Centro Regionale di Epidemiologia Veterinaria, Istituto Zooprofilattico Sperimentale delle Venezie, Legnaro, Padova, Italy
Mickle, Tom	Merial Limited Avian Global Enterprise, Duluth, Georgia, USA

Munster, Vincent	National Influenza Centre and Department of Virology, Erasmus Medical Centre, Rotterdam, The Netherlands
Ortali, Giovanni	Aia Veronesi Group, Treviso, Italy
Picault, Jean-Paul	Agence Française de Sécurité Sanitaire des Aliments, Ploufragan, France
Pittman, Maria	European Commission, Brussels, Belgium
Prandini, Francesco	Merial, Assago, Italy
Renström, Lena	National Veterinary Institute, Department of Virology, Uppsala, Sweden
Schrijver, Remco	Animal Sciences Group, Wageningen UR, Lelystad, The Netherlands
Senne, Dennis	National Veterinary Services Laboratories, Veterinary Services, Animal and Plant Health Inspection Service, US Department of Agriculture, Ames, Iowa, USA
Shortridge, Ken	Dept. of Microbiology, The University of Hong Kong, Hong Kong, China
Stegeman, Arjan	Utrecht University, Faculty of Veterinary Medicine, Department of Farm Animal Health, Utrecht, The Netherlands
Swayne, David	Southeast Poultry Research Laboratory, Agricultural Research Service, US Department of Agriculture, Athens, Georgia, USA
Valder, Wolf-Arno	Office International des Epizooties, Brussels, Belgium
Van Boven, Michiel	Animal Sciences Group, Wageningen UR, Lelystad, The Netherlands
Van den Bosch, Goossen	Intervet International B.V., Boxmeer, The Netherlands
Van der Goot, Jeanet	Central Institute for Disease Control, Wageningen UR, Lelystad, The Netherlands
Van Rooij, Eugene	Central Institute for Disease Control, Wageningen UR, Lelystad, The Netherlands
Velkers, Franscisca	Utrecht University, Utrecht, The Netherlands

Printed in the United States
41209LVS00003BB/1-3